Copyright © 2019 Kieran F. Geoghegan
All rights reserved.

Creating Cures:

A Young Scientist's First Job in American Biopharma

Kieran Geoghegan

DEDICATION

To Joanne, Barbara and James

Also by Kieran Geoghegan

Enzymes, Wizards and Secret Passages:
Intuitive Lessons in Biochemistry

Cover design by Barbara Geoghegan (www.barbaragdesigns.com)

Cover photograph: Getty Images iStock. Used under license.

CONTENTS

	Preface	i
1	For Whom the Nobel Tolls	1
2	Biopharma USA	7
3	The Agency and The Patent Office	18
4	Why You Will Be Hired	26
5	Applying and Interviewing	40
6	Joining the Team	47
7	Planning Your Work and How It Will Be Judged	57
8	Why You Might Lose Your Job	63
9	The Universal Drug Researcher	69
10	Presentations and Publications	75
11	Lifestyles of the Big and Little Companies	83
12	Academic Drug Discovery	88
13	Should I Stay or Should I Go? Moving Into Management	96
14	Science-Based Employers Beyond Biopharma	101
15	Call Me By My Brand Name (Until My Patent Expires)	105
16	Marvelous Medicines. Part 1: Biologics	113
17	Marvelous Medicines. Part 2: Small Molecules	129
18	The End of the World is Nigh (Maybe)	140
19	If It Feels Good, Do It!	150

Disclaimer

The author is neither a physician nor a veterinarian. References in this book to medications should not be taken as medical or veterinary advice of any kind. They are intended to illustrate the scientific and commercial activities of the biopharmaceutical industry for the interest of research scientists considering careers within it.

Financial Disclosure

The author receives retirement benefits from Pfizer Inc. He also holds stock in the company and in a number of other pharmaceutical companies named in this book.

Preface

Reader, is this you? You are a scientist in North America being trained in a discipline central to drug discovery such as biochemistry, molecular biology, organic chemistry, computational biology, computational chemistry or structural biology. For several years you have worked hard to gain the BS, MS or Ph.D. degree that is the next milestone on the road to your goal of living the life of a research scientist. You may even have passed those markers and jumped into the satisfying life of a postdoc, savoring the challenge of tackling part of a major problem on your own.

But it's time to think of finding a real job. You like and respect your mentors in the academic world, and there was a time when you dreamed of following in their footsteps by leading your own research group and teaching. Then you learned that opportunities to do that are far outnumbered by the good candidates available to fill them. You may not have been lucky enough with your work, or chosen a sufficiently hot area of research or a sufficiently influential mentor, to acquire the glow that surrounds a future star of academe. You are unlikely to be swept into a university, given start-up money, and handed the exciting, if strenuous, challenge of making it as a professor. Nor will one of the great research institutes sing sweet arias under your balcony.

What should you do? The great alternative is Industry, which even by its name sounds dangerously close to hard work. Its main business is marketing medicines, also called drugs (which your mother told you to avoid), and critics say that its leading companies invent new diseases and then charge too much to semi-cure them. There is also biotech, ranging from the start-ups across the street from campus to companies that started small and became giants. Taken as a single industry, which increasingly they are, pharmaceutical and biotechnology companies can collectively be called biopharma. Ironies aside, they play an important part in our society and our lives. Together they make up our capability for discovering new medicines and bringing them to patients.

Like virtually every enterprise in our global economy, biopharma operates under fierce pressure to succeed or die. Even so, its roadmap for success could also be your own, and it's important to understand it. If your goals and the industry's goals are compatible, it could be the start of a beautiful friendship.

Even as it flexes and evolves in response to new science and shifting winds in the commercial world, biopharma operates in a space of massive opportunity. People everywhere are eager for new medicines that bring medically meaningful benefits to persons affected by disease, and the foundation of our economic system is that one person's need is another's opportunity. Creating these medicines starts with scientific work that picks up the advances made by academic scientists like yourself, and tackles the seriously difficult and widely underrated task of inventing new medicines. Biopharma's foundational strategy is to target disease through the windows of insight opened by great basic research.

If this sounds exciting, and something to look into, then it should and I am glad. But take careful bearings as you steer a course into biopharma. The competition for places and success is severe and the waters can be choppy. This book shows you both sides of the story, the rough and the smooth, the joys and the risks.

We can do many things with our lives, and judging the choices that others make is not always fair or useful. But if you are convinced that by living the life of a scientist you can make a meaningful, positive mark on the world, the biopharma industry is a setting in which your hard and honest work can make that dream a reality. Let's take a look at it together.

Kieran Geoghegan – 2019

Mystic, Connecticut

For Whom the Nobel Tolls

For a pleasant read – a book nearly as good as this one – try Sir Peter Medawar's *Advice to a Young Scientist*.[1] Medawar (1915-1987), a distinguished immunologist, was warmly concerned for his fellow-scientists, and his book is a treasure chest of humane, sensible suggestions for ambitious young researchers.............of a bygone era.

Old-school phrasing aside, *Advice* remains agreeably witty and packed with evergreen wisdom about the scientific life. Sir Peter was a forward thinker. As he deals with many topics – each, delightfully, receiving no more words than his thoughts require – his ideas are often as fresh as when they first appeared in 1979. His separate chapters on Sexism and Racism in Science and the transactions of younger and older scientists are enlightened and enjoyable. His gently donnish voice insists on principle with the easy authority born of knowing that his obituarist will not overlook his Nobel Prize.

And yet, so much has changed since he wrote: as they tend to do, forty years have transformed science and the world. Computers; the internet; instant information; social media; broad globalization; and the renaissance of China, not least in research. In particular, with regard to careers, the global abundance and mobility of highly-educated workers has boosted competitive pressure in the marketplace for scientific employment. It's harder than it used to be to find a satisfying job as a scientist.

To mention this is no criticism of Medawar; it is not as if the world never changed before 1979. But the era that he spoke of is long gone, and young people considering scientific careers still need and deserve guidance. Medawar's Young Scientist – the person to whom his book is addressed – is clearly expected to work in basic research, most likely in a university or government-funded institute. His or her research topic is to be chosen carefully at the outset and then pursued for many years. That career track still opens to a fortunate few, and is no easy ride, but they represent a minority of Young Scientists seeking a path forward from academic training today.

Advice is not in short supply, and much of it comes Medawar-style from high achievers to future high achievers. For example, *Letters to a Young Scientist* by the great Harvard sociobiologist E. O. Wilson nicely resembles the guidance of a caring uncle sharing wisdom and yarns from a lifetime of pioneering research.[2] Like Medawar's, Wilson's advice comes from the scientific summit, and he urges his reader to blaze an independent trail even from teenage years. Escape the crowded fields, he says, and give yourself entirely to science. Shun vacations in favor of productive field trips to Australia or the Amazon. Being a scientist does not require genius (the really smart people become tax accountants), but a passion for personal discovery and a tough but ethical approach to competition will carry the Young Scientist to the heights.

Advice like Wilson's floats down from the pinnacle of achievement and may not always help the hiker in the foothills. Young Scientists may cringe at learning that their fields are too crowded to offer scope for more than a "steady stream of small discoveries."[3] But even if this is true in chemistry and biochemistry, which it largely is, why does it have to be the most important consideration? The discovery of a new medicine, for example, is a chain of relatively minor advances that together make a highly valued one. Forging links for that chain is something to be excited about.

Publishers pump more advice through the same pipeline. Barry Marshall, whose intrepid self-experimentation proved that a bacterial infection causes stomach ulcers,[4] wrote a children's book called *How to Win a Nobel Prize*.[5] A lecture series by the molecular biologist Michael Bishop is titled *How to Win the Nobel Prize: An Unexpected Life in Science*.[6] Peter Doherty, a prizewinner in Medicine, authored *The Beginner's Guide to Winning the Nobel Prize: Advice for Young Scientists*.[7] It seems that Stockholm is calling you, but wait! The authors of these last two books disarmingly warn that their contents are only tenuously connected to their titles – it's just that books called *The Extreme Heaviness of Being a Serious Researcher* or *Twenty Years Before the Microscope* tend to stick to bookshelves, while the prospect of learning how to win a prize is less resistible.

It's only fair to mention one book written from a perspective in which career survival replaces the Nobel Prize as a supreme objective. This is *A PhD is Not Enough!* by Peter J. Feibelman,[8] a wryly and drily humorous coaching manual for the striving young researcher. Mindful that the young investigator walks a thin high wire with a real risk of falling, Feibelman suggests commonsense strategies to be employed as the first nervous steps are taken. In a few cases, he touches on subjects that also appear here, but his background in physical science and emphasis on the academic career track give his excellent book a different flavor.

Every field has its heroes, and science is no exception. Hard work, inspiration and luck lead its champions to the summit, many of them openly surprised by their own success. For every Jim Watson, whose brilliant but darkly self-revealing memoir *The Double Helix*[9] teaches the best lesson of all about winning a Nobel Prize – *choose the right problem to solve* – many others have been nudged into the limelight by the Brownian motion of chance and circumstance. Some are authentic geniuses, like Albert Einstein or

Francis Crick, but all have had the luck to work on the right problem at the right time, even when they did not know it.

Watson knew exactly what he was doing when he chased the molecular structure of genes. Had genes been made of old boots, he would have studied old boots; it was only because Oswald Avery's group had recently shown that genes were made of DNA that he developed an interest in that molecule. The first of his several exhibitions of genius was to recognize what mattered most in 1950 in the emerging field of molecular biology. He somehow understood things about research in his early twenties that others take a lifetime to unravel. The choices that he made took uncommon vision, recklessness and ambition together with support and tolerance from senior figures. Only a few fortunate young people today find themselves positioned to follow his example.

Whether it is the Nobelists themselves (Win a Nobel!), or even the grounded Dr. Feibelman (Get a Job!), these authors take it as understood that the main goal of a career in science is personal distinction. Be problem-focused, not technology-focused, warns Feibelman, or they will make you the specialist who runs a big instrument for many projects and gets minimal credit. Find something new to work on and lead your field, says Wilson. Choose important problems, urge Medawar and Watson.

Perhaps some scientists are driven by their desire for personal recognition, but does everybody think like this? Not only in science but also in other parts of life, I was always more motivated to achieve the goals of a team than I was to make myself a star. True, I am a modest man with much to be modest about,[10] but if you share this personality, don't take it as a failing. It will also be your greatest strength if your life choices align with it. You may find it hard to compete with the individualists if you play for the same prizes, but there can still be a welcoming place for you in science. You own your own life and motives, and the only judgment that matters about your career choices is your own.

Ambition is not a crime. It's not even a mistake. And yet, winning a Nobel Prize is a gift that comes to very few. Not only are they talented and hard-working, but they also must be lucky.

Regarding luck, science is not a pure meritocracy. Olympic sprinters contest supremacy under equal conditions, but undertaking a career in science is more like joining an army. There is no draft, and everyone volunteers to begin with, but the recipe for most individual careers blends a dash of personal choice with a cup of the random. Whether this leads to success or disappointment, security or a struggle, is not a controlled process. Shockingly few scientists have anybody giving them sound personal advice based on real experience as they build a career. Therefore, find a mentor if you can, but be warned – it may take twenty years to know how well advised you were.

The same caution, of course, applies to the advice in this book. My experience will not be your experience. Too much has changed, and sixty years or more could elapse between the beginning of my career and the end of yours. Seek wisdom wherever it might be found, but always test it against the realities of your own life.

Lots of people these days offer to fill the mentorship gap, and much sincere, reasonable and worthwhile advice appears in the back pages of journals like *Nature* and *Science*.[11] Much of it comes from scientists who are themselves on the pilgrimage, the odyssey, that the scientific life becomes. First the travel tends to be physical – mobility in pursuit of opportunity is normal for aspiring researchers – and then, for those who find a permanent post, it becomes a journey of the soul. The person may remain in one place, surrounded by the dance of real life, but the scientific mind constantly migrates away into its private universe. Whatever demands this may place on the scientist, it is much harder on anyone with whom the scientist shares a life. Medawar writes sympathetically about this in *Advice to a Young Scientist*.

If you receive the necessary opportunities for development, early success and supportive patronage, a career as a professor still offers great attractions. To lead a research group, speak and publish with relative freedom, and send students forward on careers of their own is an exciting privilege. Admittedly, it comes at the cost of enormous effort, a huge trial in the form of the tenure system, and a struggle for research funding that lasts as long as your career.

There are not enough tenure-track academic opportunities in US colleges and universities to accommodate everyone who completes a Ph.D. in bioscience and a productive postdoctoral stint.[12] For those excluded by selection or preference, a major alternative is a career in the biotechnology and pharmaceutical industry – biopharma for short.

Just as importantly, many true lovers of science who end their academic journeys at the BS or MS level also require employment. Whatever the scientific level, the huge, complex, fast-changing and challenging biopharma industry offers vast opportunities for bioscientists to realize what should always be their supreme ambition – to live the life of a scientist.

Biopharma USA 2

Science is interested in the entire natural world, but biomedicine receives more than a fair share of its attention. The United States Government, for example, has recently spent well over thirty thousand million dollars per year on its biomedical research and granting establishment, the National Institutes of Health (NIH).[13] That's about four times as much[14] as it sends to the National Science Foundation, which manages and funds a wide range of important research with massive economic relevance. And, research aside, why do healthcare-related expenditures account for about one in every six dollars spent in the US economy?[15]

Jim Watson recently pointed out that Congressional lawmakers and their families are as likely as anybody else to be affected by illness, so that NIH funding always finds strong support.[6] Every human society recognizes the value of medicines, beginning with folk tradition that says a broth made from certain leaves reduces this kind of pain. In modern times, medicines have come from a mixture of educated accident (some antibiotics) and science (newer cancer treatments). Individuals prize good health and will pay, preferably with the protection of insurance, to recover or keep it. For governments, public health is a supreme social good because it promotes economic well-being and drives general happiness.

My brother, a surgeon, says that "surgery is brilliant if you operate on the right thing." I believe it, but surgery and hospitalizations are distressingly expensive and need to be reserved for when they are really needed.

Huge value therefore lies in medicines that *keep people out* of hospitals and nursing homes. Data from the Centers for Medicare and Medicaid Services show that the $329 billion spent in 2016 on prescription drugs by all payers represented a modest 11.6% of US spending on personal health care (outlays for goods and services relating directly to patient care).[17] Now, some drugs are expensive by any standard, costing over $100,000 per year, but biopharma argues that these costs represent good value to the health care system.[18] This is by comparison with the higher costs of treating people who fall deeper into illness, and as support for further expensive and risky research. The second leg of this argument obviously relates to scientific employment. No funding, no fun!

To interpret this picture and appreciate the challenges facing the industry that we hope to enter, we need to understand where new drugs come from. In economically developed countries, the system for providing innovative medicines stands on two pillars.

The first pillar is the academic sector, extensively funded by Government in the US and charged with conducting and publishing research in basic science. Alongside it are the very significant contributions of charitable foundations, of which the Howard Hughes Medical Institute may be the most prominent in biomedicine. The value set on different parts of this work is determined by peer review related to publication and the allocation of funding through grants. Quite generally, the highest value will be placed on research that uncovers fundamental mechanisms and systems in biology that ultimately, even if not immediately, relate to human health.

The second pillar is biopharma, the privately-funded industrial sector. Biopharma sustains itself by turning insights from basic research into new medicines whose value is determined by the medical marketplace (Figure 2-1). Nobody has to prescribe or take a newly marketed medicine: its value is determined in the real world by the benefit that it delivers to patients compared to existing therapies. This creates huge selective pressure on the

Chapter 2: Biopharma USA

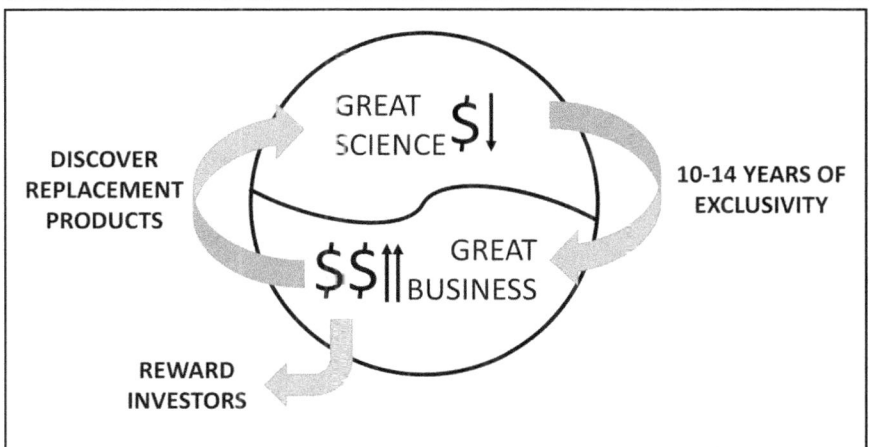

Figure 2-1. The current business model of established biopharma companies. Ideally, a fraction of the revenues from presently marketed medicines funds the discovery of future medicines to sustain the cycle.

company that brings it to market. A drug discovery program cannot be based on its perceived elegance or scientific interest: it must, above all, be intended to result in a new medicine that outperforms whatever is currently available and commands a market that makes its discovery and development profitable. That is a high level of challenge.

Pricing is an important part of the conversation, and it is sobering that our Young Scientist's career can be massively affected by a factor completely beyond his or her control. Medicines are largely paid for by private or governmental insurers for whom cost control is a high priority, and it is clear that they will not embrace an expensive new medicine unless it delivers clear medical value and, preferably, net cost-savings. And even if drug treatment clearly offers savings over an expensive non-drug alternative, the system will not automatically pivot in response.[19]

Biopharma's strategy for profitability is under stress and its successful continuation is not a certainty.[20] People around the world welcome medical innovation, but the US is exceptional in preserving a market in which the prices of medicines are not entirely regulated. Other advanced countries, such as Canada, the United Kingdom and Germany, maintain health systems largely operated by their governments, and in which government agencies negotiate drug prices with biopharma companies. This can lead to patients being denied access to innovative medicines if the government's analysis deems them insufficiently cost-effective (or simply unaffordable) and the supplier cannot lower the price to a level that the buyer accepts. Anguish and political ramifications naturally ensue from operating this process. Even though it is an unavoidable consequence of resources not being infinite, patients and their families naturally suffer when a source of medical hope is denied. The same end-point is often reached in the United States, except that it is the limitations or lack of health insurance policies that lead to denial of access. In the US, therefore, denial more often occurs in a disconnected set of individual cases rather than in the context of a public decision that affects everyone in a patient group.

Recent cases in point overseas have included debates in Ireland over the availability of ground-breaking new drugs to treat cystic fibrosis,[21] and reviews of recently developed cancer drugs by the National Institute for Health and Care Excellence (NICE), which assesses the cost-to-benefit ratios of drugs for parts of the United Kingdom's National Health Service.[22]

This book is not about the complicated politics of drug pricing, but our main interest – the career prospects of scientists – is closely tied to the business prospects of biopharma companies. Innovation is widely viewed as the key to medical progress and society's ability to afford health care. Without profitable innovation that permits research-based businesses to refresh their resources, the companies seeking it will decline. When that happens, jobs for scientists will melt away.

Chapter 2: Biopharma USA

Scientists working in each sector accept risks and strive for rewards. On the academic side, unproductive scientists or teams quickly lose the opportunity to engage in research as grants are sent elsewhere. For research leaders, the traditional rewards for success are prestige, recognition and expanded research opportunities. Recent decades have added the excitement and financial incentive of entrepreneurship in biotechnology, an interesting ramification of academic life that we will discuss later.

On the industrial side, a research-based company's inability to achieve profitable innovation leads to business failure, abandonment by investors, and closure or being taken over by a competitor. Takeovers are disruptive at best for the business and often worse for staff. Success, on the other hand, leads to rewards in the marketplace and the continued and increased availability of funds for new research, with rewards going to the participants on a scale determined by their level of responsibility.

Just as in finance, technology and any business subject to disruption by innovation, the stakes are high on all sides of the biopharmaceutical enterprise. The young research scientist who considers a career in the biopharma industry, now more than ever, must enter this arena with open eyes and educated about the potential for future change, both positive and negative.

The worldwide drive for medical progress is held together by the ethical values of physicians and scientists. Scientific publication places basic advances in the global public domain, but most countries also accept the value of certifying and protecting intellectual property through patents. Together with careful governmental regulation of market access, the patent system is an indispensable pillar of innovation that the would-be company researcher must understand (see Chapter 3).

Just a few countries contain both parts of the innovation system at a world-class level. In the US, for example, world-renowned universities combine with NIH and private institutes to make up a mighty capacity for basic research, while an array of biopharma companies ranging from huge to tiny forms an imposing industrial sector. With lower numbers but equal quality, the UK, Switzerland and Germany are also home to large-scale enterprises of both types as well as their own biotech sectors. China is moving to join them, while the industrial sector in India, itself a great scientific nation, has been more oriented toward the provision of affordable versions of existing drugs. But human health is a universal concern, and many countries advance it through high-level academic research on a proportionate scale. Not only do they see this work as a duty to the global community and part of being an advanced nation, but they often aspire to seeing profitable biopharma companies spawned from their research labs. Successful drug companies are found in some smaller countries, such as Copenhagen-based Novo Nordisk, but biopharma is truly global and successful companies that begin in smaller countries are often absorbed by a multinational giant.

Assuming that our potential employers are research-based biopharma companies, it is a good idea to appreciate the commercial scale of the larger ones relative to well-known businesses. In 2018, for example, the three largest US-based biopharma companies by global revenues (sales) were Johnson and Johnson ($82 billion), Pfizer ($54 billion) and Merck and Co. ($42 billion).[23] Comparable companies included Procter and Gamble ($67 billion), Disney ($59 billion) and Coca-Cola ($32 billion). Clearly, we are dealing with large companies at the top end of the scale.

A virtual continuum follows of variously-sized research-based companies, all the way down to small-scale start-ups. All of these represent potential places for the Young Scientist to work. Many aim to discover new medicines and bring them to market, perhaps with the help of partners, but an allied sector is formed by small companies that provide specialized technical support to the true biopharma enterprises. For example, the La Jolla-based company ActivX Biosciences Inc., a subsidiary of Kyorin Pharmaceutical Co. (Tokyo, Japan), maintains useful assays based on chemical probes that discovery teams can use to evaluate the specificity of their lead compounds among targets of a certain class.

To consider only US-headquartered companies as employers would miss part of the picture. The vibrant American scientific scene and the relative ease of starting and – when necessary – terminating business activity leads many foreign companies to favor the US as a research location. Scientific hubs in which academic and industrial research run side by side have emerged in several centers, with companies strongly drawn to the skilled staff, useful partnerships and academic contacts that they offer. The most prominent are the San Francisco Bay Area and the Boston-Cambridge metropolitan area, but the Research Triangle in North Carolina, the San Diego area and the Montgomery County region northwest of Washington DC are among others that have gained real significance in bioscience. Virtually all the companies active in these and other hubs advertise both their activities and their peerless corporate virtue on nicely designed web sites that include lists of open positions.

Table 2-1 lists some larger companies with their global sales for 2018 and the US states that are homes to their research facilities. Corporate cultures and fortunes vary, of course, as do the types of work done at different research centers. The length of this list and the financial power of the companies that comprise it should persuade emerging scientists that the industry is strong, but they must also remember that many well-trained scientists aspire to work in it or to continue doing so. Rather like professional baseball or the Tour de France, the everyday level of effort and competition is strenuous and requires serious dedication to keep pace.

Large companies, with occasional exceptions, are publicly owned, which means that their ownership is divided into a large number of shares traded in the public stock markets. Galloping alongside established companies, especially in the celebrated biotech hubs, is a virtual stampede of small companies. These tend initially to belong to individuals or private investment (venture capital) groups that detect promise in certain ideas and provide seed money to begin the work. If early efforts prosper, more private investment can follow. Then, if enough progress can be claimed, the time may come to offer shares in the company to the public market. This process, called an initial public offering, is a way for owners or investors who held large stakes in the company at the outset to reap a financial reward while, at the same time, attracting substantial funding for the company's future growth.

Another much-admired feature of the American economic scene is its ability to allow promising new ideas to be tried out. No stigma is attached to people whose honest effort with a smart idea ends in failure; they are welcome to step up and try again if their next brainwave can attract investors. Silicon Valley showed the way in this regard, and biotechnology has followed its lead. The supply of venture capital available to fund these experiments waxes and wanes with the economy, but some bullish forecasters anticipate that emerging methods grouped under the banner of personalized medicine will thoroughly disrupt the medicines business. Together with this prediction comes the suggestion – for some, the ambition – that the revenues of today's leading companies can, before long, be diverted to organizations that better master the future.

This is another package of issues for the Young Scientist to survey. On one hand, if you enter a more-or-less traditional drug company, will the business be undercut by radical innovation just when you have dug in and become committed? On the other, what

Chapter 2: Biopharma USA

Table 2-1. Selected Major Research-Based Biopharma Companies Conducting Research in the United States

US-Based Companies			Foreign-Based Companies with US-Based Research Facilities		
Company	Revenues (M$)	US Research (Major)	Company	Revenues (M$)	US Research (Major)
Johnson & Johnson	81,581	NJ, PA, CA	Roche (Genentech)	57,199	CA, MA, IN, NY
Pfizer	53,647	CT, MA, MO, CA	Novartis	51,900	MA, NJ, CA
Merck	42,294	NJ, PA, MA, CA	GlaxoSmithKline	39,920	PA, MA, NC, CA
AbbVie	32,733	IL, MA, CA	Sanofi	39,149	MA, PA
Eli Lilly	24,556	IN, MA, CA	AstraZeneca	21,049	MA, MD
Amgen	23,747	MA, CO, CA	Boehringer Ingelheim	19,587	CT
Bristol-Myers Squibb	22,561	MA, CT, NJ	Novo Nordisk	17,012	IN, WA
Gilead Sciences	22,127	CA	Bayer	16,674	MA, CA
Celgene*	15,281	PA	Takeda	15,965	MA, CA
Biogen	13,453	MA	Astellas	11,783	CA
Regeneron Pharma.	6,711	NY	Otsuka	11,690	MD
Zoetis	5,825	MI	Daiichi Sankyo	8,381	CA
Alexion Pharmaceuticals	4,130	CT, MA	Merck KGaA**	7,007	MA

Revenues are worldwide for 2018 in M$ (millions of dollars); source, company reports; currency conversions by the author

*In January 2019, Celgene agreed to be acquired by Bristol-Myers Squibb for $74 billion.   **Healthcare revenues only (Merck KGaA).

opportunities can you detect for your own career in the progressive new areas of genome-based diagnostics, micro RNA's, CRISPR/Cas9-based gene editing, or whatever it may be? You must keep your head up and be aware of these breaking waves, because the leadership of any company that you work for will certainly be open to the new opportunities that they bring. Be the beneficiary of change, not its victim.

Many small biotechs are privately held companies that must, of course, obey the laws of the land, but are relatively free from obligations to disclose their financial status. When a company "goes public," everything changes and the fairness of markets trading the company's stock is intended to be protected by requiring the company to disclose developments that affect its value. Among these could be the success or failure of a clinical trial, an important arrival or departure in management, a significant investment in new facilities, or a new research collaboration. Regular public financial reports, usually given quarterly, are a key requirement that allow investors to keep an eagle eye on the fortunes of the company that they partly own. If they don't like what they see, they can go to the market and seek a buyer to take the investment off their hands. When news about a company is bad, the price of shares in its ownership falls. Its total value, or market capitalization, is calculated as the price of one share multiplied by the number of shares issued. Online stock quotes generally include a "market cap" value.

There are too many small biotech companies, and their fortunes and names change too quickly, for a list given in Chapter 18 to be more than a snapshot. A search engine will turn up many more. Again, most companies maintain a web site intended, among other things, to interest potential employees. We will explore the special functions and purposes of small companies in a later chapter. Some are trying out daring new concepts in drug discovery, perhaps with the aim of being acquired by a bigger company;

some, like ActivX, may specialize in providing special technologies that catalyze discovery efforts elsewhere.

The Young Scientist's range of potential employers stretches from corporations with workforces in the tens of thousands down to companies with staff numbers in single digits. Personal preference, location, opportunity and family factors all contribute to determining which type of employer a scientist will approach.

The Agency and the Patent Office 3

The United States Food and Drug Administration (FDA) is the government authority that evaluates and licenses medicines for sale in the United States. In the world of mainstream medicines of all kinds, including vaccines, its word is absolute law and well-advised elements of the industry defer to it with maximum respect.

To be approved for sale by FDA, new medicines must be deemed *safe and effective*.[24] This simple phrase is heavy with significance, and reflects the physician's Hippocratic oath to "above all, do no harm." This priority dictates that Phase 1 of the clinical trials of a new medicine – the first study in which it is given to human patients – is mainly about finding out if it is safe. Only when it has been shown that the would-be medicine causes no unexpected toxicity does the trial process move to Phase 2, when it is tested for signs of efficacy against a certain illness in a small group of patients. Encouraging results at that stage lead to Phase 3, in which the trial sponsor – usually the innovating company – seeks definitive data that will support a New Drug Application (NDA) for clearance to market the medicine. After taking advice from medical experts and other interested parties, often including patients and their advocates, FDA reviews the NDA and issues a ruling to license the medicine or deny the application.

FDA's crucial role as a regulator does not preclude it from acting as a creative, positive force in the inception of new medicines. An historical example is the genesis of the statin class of cholesterol-lowering drugs.[25] Currently, as 2020 approaches, FDA is actively mapping out the future path of gene therapy.[26] FDA's own physicians and scientists can assess independently what is required in new medicines for certain conditions and what has to be avoided, and share their ideas with innovative companies. Its mandate is to weigh the benefit of approving a new medicine against the risk, guarding against mishaps while also recognizing that life-saving and life-extending medicines are urgently needed by many citizens. Just as many physiological systems function by achieving a balance between opposing forces, the FDA constantly tries to optimize the risk-benefit ratio in each case that it considers.

From our aspiring researcher's perspective, the chemistry and biology that go into making a new medicine happen largely before it enters clinical trials. Even work related to developing means of following drug action, such as creating the assay for a biomarker, takes place in the preclinical space. Therefore, we will not examine the clinical phase in greater detail in this chapter, except to note that it requires specialized professional expertise quite different from the skills needed in laboratory work. With suitable training, many former researchers choose to develop their careers by transitioning into these roles. Compared to preclinical work, the clinical phase is long, expensive, complex, and heavy on organizational and administrative effort.

Only licensed physicians can supervise the administration of medicines to volunteers and patients, so they will always be the people in charge of what happens during clinical trials. If your training is purely scientific, be aware of this immovable stratification of responsibility and prestige. Many stimulating and well-compensated employment opportunities exist for scientists in the clinical phase, but they will always be under medical leadership.

National and international regulators elsewhere in the world govern the admission of medicines to their own markets, and do not always reach the same conclusions as FDA. For example, biopharma companies wishing to market a medicine in countries of the European Union submit them to review by the multinational

European Medicines Agency (EMA). In Australia, the Therapeutic Goods Administration is the reviewing body, and so on. Generally speaking, data presented for review by these regulators are assembled by companies with the utmost care. Medical writing – the production of formal reports pertaining to clinical trials – is a professional specialty in itself, and skilled practitioners with experience are in high demand.[27]

Together with the need to satisfy regulators, a second prominent reality for research-based biopharma companies is the urgency of claiming and securing inventorship for prospective future products. FDA and its international peers are distant considerations for our hands-on researcher in early-stage drug discovery, but the urgency of securing the company's intellectual property profoundly affects everyday life in the lab. Every researcher should understand what patents are, the rights that they provide to their owners, how they are secured, and why they are crucially important.

The patent system is a piece of genius that goes back to ancient times and entered US law in 1790. Its basis is a trade-off between an inventor and the rest of society. In exchange for disclosing how an invention is practiced, the inventor receives the right to exclude anyone else from making, using and (especially) profiting from the invention for a fixed period of time. In the United States, that period is twenty years from the date on which the patent application is filed with the United States Patent and Trademark Office (USPTO). Review of the application by USPTO typically takes well over a year, after which the Patent Office takes its first action to decide whether all, some, or any of the patent's claims will be "allowed." The process has complications – the application can be amended, outsiders can object to the Patent Office's rulings – but the inventor's hope is that a patent protecting the claimed invention will eventually be issued.

The public benefit of the patent system is to guarantee inventors

Chapter 3: The Agency and the Patent Office

who have used their time and money to bring something new and useful into the world an opportunity to profit from their initiative. Until the term of the patent has expired, others have to make an arrangement with the inventor if they wish to use the invention. Anyone can patent an improved form of the invention at any time, but the original claims remain the intellectual property of the inventor through the patent's term, and the inventor can call anyone who infringes on that right before a judge and demand that the patent be upheld by authority of the court.

To be patentable, the invention has to be new, non-obvious and in some sense useful. Failure to pass even one of these filters would be grounds for rejection.

What can be patented? Machines, processes, circuits, products of manufacturing, even genetically engineered animals and plants – but the most important type of invention to biopharma and its scientists is a "composition of matter." This phrase describes a specific chemical or biochemical such as a potential drug, and may also include details of how it is formulated. With a properly characterized small molecule, there is no ambiguity about the structure, and inventors who bundle a structure with evidence that it is useful for treating a certain disease have a strong position if the Patent Office judges that their invention is new. But, if they or somebody else has previously put this information in the public domain either as a patent claim or any other form of disclosure, then the examiner will rule that "prior art" exists and the application will be rejected as lacking novelty. With a biological molecule such as an antibody, where intrinsic variability (like glycosylation) makes molecular characterization less tightly defined, a composition of matter claim is still possible based on amino-acid sequence and other properties.

The goal of early-stage drug discovery is to select a specific molecule with potential to become a marketed drug. This process typically consists of examining very many possible structures, characterizing their activities and properties, and narrowing the focus to a single option. However, a feature of the patent system is that patent applications are published eighteen months after being filed, so that competitors get an early look at the structure envisioned as the future drug. They can then try to leapfrog the claimed invention by discovering something even better and filing their own claims. Partly for this reason, patent applications usually describe entire classes of related molecules (small or large), among which the actual preferred candidate lurks as one face in a crowd. In this and many other ways, companies obey the rules of the game while still seeking every possible competitive advantage.

The other reason for including related structures is to lay claim to more than the precise structure for which hopes are highest. Quite fairly, this prevents competitors from making trivial modifications to the active compound – adding an insignificant methyl group to a small molecule or changing an unimportant amino acid in an antibody sequence – and then claiming the slightly altered molecule as novel. But if the claims are unreasonably broad, the examiner will perceive that the invention has not been adequately limited. To claim everything is to claim nothing in particular. The result of this calculus is that most patent applications list a cascade of claims that begin with the most specific and important ones, and then broaden to form an aggressively inclusive list. The inventors will hope that the essential claims at the top of the list are granted, and take their chances on the examiner's willingness to allow more expansive claims.

That is a simple version of the story. In the real world, activity around patents can be very complicated. Outside the US, a United Nations agency known as the World Intellectual Property Organization (WIPO) performs some of the functions performed within the US by USPTO. Applications are published by WIPO, but actual patent rights are still granted by individual countries or

Chapter 3: The Agency and the Patent Office

by regional offices to which they delegate this duty.[28] Through international treaties, important rights are exchanged between countries. For example, USPTO may honor the filing (priority) date associated with a patent application to WIPO, and vice versa.

Biopharma companies naturally want to obtain exclusive rights to their inventions in as much of the world as possible, and managing the global patent strategy for a candidate medicine is a vital, well-compensated function in any organization. Because the system involves legal disputes that can make or break companies, attorneys and other legal staff are an essential part of this apparatus.

These legal experts would not have jobs, however, if it were not for the successful efforts of our own heroes, the discovery scientists who work in laboratories. The patent system requires individuals – not companies – to be named as the inventors on any patent, and chemists traditionally regard it as a huge accomplishment to be named as an inventor. The rights associated with the patent are then typically assigned by inventors to their employer.

The requirement for patent claims to be novel strongly influences the ability of research scientists in industry to publish their work. As we will see later, companies generally encourage publication subject to the strong proviso that the needs of the company to establish and protect its intellectual property are respected. For this reason, scientific publication is carefully controlled by companies and commonly has to wait for patent filings to be secured.

A more immediate effect for the working scientist is the absolute need for immaculate record-keeping of research work. In the case of a dispute about patent claims, companies may need to produce the original records of their research. These need to be signed and witnessed with dates included, and companies have increasingly turned to electronic notebook systems that collect the information in a secure archive. Maintaining the e-notebook is an important duty of the lab scientist in a company, and neglecting it can have dire consequences for both company and individual. A good lab supervisor takes care that this never happens on his or her watch.

Bringing a drug to market, then, requires success in two channels of effort. Our company must securely own exclusive rights to the relevant composition of matter, namely one or multiple patents that describe the active formula and additional specific features of how it is administered for one or more stated medical purposes. We must also have received a letter of approval from FDA to market the medicine in the United States. Patent protection and regulatory approval throughout the world will also be sought at the same time.

Now we can market the medicine, using a hard-working sales force to brief physicians on its benefits and risks, but only in the specific medical areas for which it is approved. "Off-label" promotion is illegal and can have disastrous consequences. We can also advertise in medical journals, where the detailed information that prescribers require can accompany the promotion. More controversially, the United States permits direct-to-consumer advertising such as the ubiquitous commercials for branded medicines seen on television network and cable news shows.

If no medical mishaps are detected when the medicine reaches a wider population of patients, the FDA's green light for marketing remains on indefinitely. However, *exclusive* rights to market the medicine have a finite lifetime that generally coincides with the remaining validity of the relevant patent.

As with other aspects of patent management, there are subtleties. FDA can extend that period of exclusivity (even though the patent's lifetime is unchanged) by six months if the company performs studies of the medicine's active ingredient in children and submits the results for review. Companies marketing a successful drug generally have much to gain by expending the effort needed to take advantage of this rule, which is called the Pediatric Exclusivity Provision.

When exclusivity lapses, approved generic forms of the medicine can be marketed. Again, incentives are present in the system to encourage the wide availability of medicines at reasonable cost. The first approved generic form of a medicine itself receives six months of exclusive rights to be sold as an alternative to the branded product, the owner of which may also introduce a generic form of the product during this time. After that initial six-month window, all approved generic forms can be sold alongside the branded original.

When the original branded drug has earned large annual sales, manufacturers of generic medicines have a strong incentive to enter the market and slice off a piece of the action. When a number of them do so, the price of the medicine tends to fall dramatically, and the system works as intended. Greater difficulty has arisen in cases of drugs for rare conditions, where the total market is modest and competition is lacking. In occasional cases of this type, cynical opportunists have abused the medical need of patients by seizing commercial control of a medicine and raising the price by a large factor. This kind of activity has energized biopharma's critics, and the fact that it is deplored across the industry itself has not spared the industry reputational damage.

Why You Will Be Hired

4

The words of Annie Dillard – "how we spend our days is, of course, how we spend our lives" – ring especially true when most of life is in the rear-view mirror.[29] Days have turned into months and months have turned into years, and the years have slipped by. If you can look back and say that you spent those days and years and decades doing exactly what you always wanted to do, you have been fortunate.

If what you always wanted included trying to make the world better and serving others, then you may have spent your time well. Consciously or unconsciously, many aspiring scientists are striving to reach this viewpoint.

Idealism is required. A research career demands a sense of purpose beyond being able to afford a home and join in supporting a family. Most bioscientists put on the white coat because they hear a call to make the world a better place by improving human health. Biopharma companies are the source, with very few exceptions, of all new medicines that come into the world. Therefore, working in biopharma offers an accessible way in which many young scientists can make their dream come true.

The connection between an employer and a new employee is always highly individual. Scientists bring their personal character and experience, and we have seen how biopharma companies range from huge and established to new and tiny. Despite this diversity, there are general and sensible reasons why working in the industry could be a good choice. To borrow from the industry's TV commercials, you should ask if biopharma is right for you.

Chapter 4: Why You Will Be Hired

Abraham Maslow's famous pyramid of needs,[30] with personal safety and security at its base and self-actualization (living your dream) at its apex, relates perfectly to employment in our market economy. The system sorts people according to their personal abilities and the work that they have been prepared for, leaving some in unhappy difficulty but offering many an acceptable way forward. Researchers knocking on biopharma's door carry the hope that their hard-won skills will fit well with a company's need to convert discovery and innovation into profit and medical advance.

Along with the ideals of living the life of a scientist and helping to discover new medicines, it is reasonable to aspire to some personal rewards. These normally include having a satisfying career as a professional, enjoying good conditions and compensation, and joining a great team doing worthwhile work. Again, these highly satisfying goals can be reached in the biopharma industry. Its great companies are *great* companies that contribute something wonderful to the world whenever they succeed in introducing a new medicine. To have a realistic chance of achieving that – it is exceptionally hard – they must provide their staff with excellent working conditions, compensation that allows them to focus on their demanding work rather than the leaky faucet at home, and a reasonable degree of stability.

But, you say, I read news stories and social media posts all the time claiming that the drug industry is "all about profits," that it is a price-gouging, disease-inventing conspiracy that fleeces people by over-charging them for medicines that the government created at NIH or paid for in universities. Why would I want a career with those bad people? And, now that I think of it, why will they hire me to do research if their business is based on ripping off other people's nearly-completed projects?

As for the canard (that's a rumor, or urban myth) that all medicines are invented by government researchers, this idea is false and comes from widespread failure to appreciate what it takes to create an approvable medicine. More on this later.

Criticizing a company for trying to make a profit is like criticizing a dog for barking: in either case, it's in their nature. Biopharma's model for sustaining itself requires the profit from one medicine to fuel the research that leads to the next, while also paying a dividend to company owners who risked their capital on a company that habitually devotes 14% or more of its sales (not just its profits) to new research with no guarantee of success. And remember, companies have only limited time in which to recover the costs of the research behind new medicines, as well as the costs of their unsuccessful efforts. Patent protection that gives the inventor company exclusive marketing rights to a new medicine often expires 12-15 years after marketing approval is gained. After that, other companies that spend no money on discovery research can produce a generic form of the medicine. When this happens with important small-molecule drugs, the price falls by a large factor and the medicine can be expected to be relatively cheaply available for as long as it is useful.

This, at least, is the theory of how the system is supposed to work. Recent attempts at price-gouging by get-rich-quick merchants have disgraced the industry, but have been rare. Also, with regard to the generic forms of biologic drugs, known as "biosimilars," it remains to be seen whether pricing follows the pattern established with small molecule agents. Because antibodies and other protein drugs emerged later and are characterized less definitively than small molecules, the rules for establishing that generic copies of previously approved agents are functionally equivalent to them were slower to be agreed. As a result, biosimilars of major biologics have only recently reached the marketplace and price-lowering competition has, to this point, been less vigorous than policy-makers hoped.

Chapter 4: Why You Will Be Hired

Academic research is designed to focus on new discovery intended to open new horizons for medicine. It remains quite successful in doing so, although the recent blossoming of "discovery-style" research modeled on the Human Genome Project – that is, programs intended to capture the entirety of information about aspects of systems – has tended to result in the appearance of massive research productivity that is not matched by immediate equivalent usefulness.

To express this another way, traditional research tended to focus on a particular system chosen for its perceived interest and significance, with good likelihood that the work would point forward to useful applied research based on its findings. The contemporary practice of collecting all the information of a certain type about a system (e.g. all the sites of protein phosphorylation in a proteome, or all the types of glycosylation found on proteins in a tissue) creates reference resources for applied research, but comes with no guarantee that any part of the work will ever be useful.

Despite this mild reservation, we continue to live in scientifically exciting times. We are currently at the dawn of the post-genomic era, with increasingly rapid genomic sequencing opening the possibility of surveying large collections of human genomes with a view to understanding (for example) the genetic bases of major forms of cancer. In structural biology, dominated for seventy years by X-ray crystallography, cryoelectron microscopy with high-resolution detectors has abruptly created an upheaval.[31] Research in proteomics has opened a new perspective on cellular function, with outstanding researchers using phenomenal mass spectrometers and clever algorithms to mine its potential.[32] Across the board, biopharma companies see new academic science as the key to new advances in drug discovery, and see the scientists involved in that work as desirable employees who can bring in the required skills. But – and the point is crucial – progress in academic science can suggest the existence of new opportunities for drug discovery, but leaves massive amounts of difficult work to be done by scientists in the industry. Your own opportunity and challenge may lie in one of those areas of work.

Biopharma's traditional thinking is that new science coupled with established drug-making expertise – something found only in industry – will be the fountain of youth for its product portfolio. Large parts of it retain faith in this paradigm, despite signs elsewhere that it is wavering. The alternative strategy is to pull back spending on early-stage research, and instead use current revenues to buy in the results of early-stage discovery conducted elsewhere. That is most often done in smaller biotech companies (which may be acquired), and occasionally in academic or government labs.

The last aspect may change as certain early-stage ventures find their way to success. For example, some companies, urgent to leave no stone unturned, have placed outreach units in major academic centers to seek productive partnerships with institutional researchers located there.

An example is Pfizer's Centers for Therapeutic Innovation at Boston, San Diego, San Francisco and New York which currently (2019) partner with over thirty different institutions to explore promising proposals from their researchers.

A different model, equally innovative, is the California Institute for Biomedical Research (Calibr) in La Jolla, at which multiple participating companies, foundations and institutions sponsor joint research programs related to disease mechanisms.

Yet another is the provision by an established company of incubator facilities in which entrepreneurs can rent laboratory space and be advised while making a start on their project. A prime example is the JLABS concept implemented by Johnson & Johnson at eleven sites in North America, Europe and China.

Urgency can be expected at any small company in which a single project is the key to the future, but larger biopharma companies – the ones with an existing profitable business to sustain – are equally hungry for innovation. This is why they are eager to embrace new ways of doing things and quick to admit that they hold no exclusive right to have good ideas. The Young Scientist looking for his or her opportunity to join the industry and participate in its research needs to be aware that things are changing quickly. It was my good fortune to spend 32 years working at research in a great company, but that gift of stability and continuity will be harder to access in future years.

With all that said, the industry continues to perform internal research. So what will get you hired? The answer depends on the level at which you apply.

The over-production of Ph.D.-level scientists by the academic sector, and moderation of its funding growth, has tended to lower the job level at which Ph.D.'s are hired by industry. Twenty years ago, they could expect to be recruited as group leaders, but today they often enter companies as "individual contributors." This means that they will engage directly in research without supervising other staff. These are not newly-graduated Ph.D.'s, but most often come in with postdoctoral experience. Publications are a must – there do not have to be many, but at least one or two should be first-author contributions in major journals – and the job interview (see Chapter 5) will always involve presenting a seminar about that work.

MS- and BS-level staff are essential in industry which, in my opinion, often grants them more respect and opportunity for career advancement than they receive in academic labs. They also will enter as individual contributors, joining a group headed by a senior scientist and helping to deliver the hands-on research required for collective success. Specific technical skills will definitely be important in determining whether a candidate is hired, and it can be a matter of luck as to whether your skills are well-matched to the requirements of the company.

Whatever the level of the position being filled, the candidate's personality and apparent potential (a judgment call) are very important. A defining feature of the industrial environment is a sense of common purpose among colleagues. Being full of scientists, a research group may be liberal in many ways, but one thing that you cannot do in a company with a healthy culture is tell a colleague that you are not interested in their problem and you are not going to help them. Therefore, any candidate is assessed for his or her potential to be a good team player and collaborator.

Regarding potential, the rapid ongoing evolution of science means that the skills that a new hire brings to a company today may be superseded within five years. You are being hired partly for what you know, but just as much because you have proven yourself as somebody who can learn. For a working scientist, the learning should never stop.

Here is an incomplete list of the hands-on research roles available to research scientists in drug discovery and development. Naturally, specific opportunities will vary between companies, and these are generic descriptions. Every one of them begins from a solid grounding in the fundamentals of relevant sciences, and then emerges as a specialty. Also, because colleagues need to understand each other's challenges, timelines and capabilities, a person working in any one of these areas needs to have a decent understanding of all the others. The biopharma industry is a rich learning environment.

Medicinal chemist (compound design). Most often a relatively senior role in which experience is a major asset.[33] Uses data from literature reports or early tests, such as high-throughput screening, to envision novel chemical structures that may possess the activity desired in a new drug. Must be able to factor in the likely metabolic fate of a compound, to anticipate and avoid genetic or acute toxicity, and to achieve patentable novelty in an area in which the chemists of other companies are also working. A central generator of the company's future intellectual property who can aspire to becoming a named inventor of a new medicine.

Medicinal chemist (compound synthesis). May work at a "junior" level (e.g. recent Master's degree or Ph.D.), but senior-level opportunities also exist for lab and group leadership. Can also contribute to compound design. Skilled at lab-scale production of test quantities of compounds envisioned by the design chemists. Has wide command of named synthetic reactions and methods, and uses chromatography for compound purification when needed. Applies NMR and mass spectrometry to determine structure and purity of products. Active compounds are the industry's ultimate products, so the ability to make them for initial testing is a central asset. Needs to enjoy being a team player who takes pride in collective achievement.

NMR (nuclear magnetic resonance) spectroscopist. May work in several areas. (1) Advanced structural analysis, elucidating fine details (e.g. stereochemistry) of products of chemical synthesis when this cannot be done by the synthetic chemist. (2) Uses specialized NMR techniques such as saturation transfer experiments to elucidate details of compound binding to protein targets. (3) Uses solid-state NMR to distinguish between different polymorphs of solid drug substance.

Process chemist. When a compound progresses to clinical trials with potential to reach the market, its chemical synthesis must be scaled up using methods that are safe, environmentally acceptable, and economically viable. This can require inventing an entirely new synthetic route. Biocatalysis will be used when possible.

Biologist – cells, assays, screens. Cell biologist familiar with techniques for culturing, harvesting and engineering eukaryotic cell lines to construct platforms in which specific functions can be assayed. Will be acquainted with transient transfection to express recombinant genes, frequently employing fluorescence-based readouts based on green fluorescent protein and its variants. May specialize in cell types relevant to a certain therapeutic area, such as liver cells or cells of the central nervous system (astrocytes or neurons). Provides team with cellular models that emulate the clinical condition to be treated, and in which the effects of tested compounds can be gauged efficiently.

Biologist – preclinical models. Usually a researcher specialized in a certain field, such as immunology or cardiovascular and metabolic disease. Trained in use of preclinical models of human disease. Competent to design and execute experiments that move tests of promising compounds forward from the cellular level. An important role for scientists trained to Master's level and beyond, but technical staff skilled in this area are also particularly valued and are not as plentiful as those with some of the other skill-sets mentioned here. Technically oriented workers may find opportunity in working with veterinary staff to maintain research capability.

Protein expression expert. Molecular and cell biologist who designs and arranges for the construction of vectors used for recombinant expression of research-relevant proteins in bacteria, yeast, insect and mammalian cells as required. Thinks forward so that protein is expressed in a format (e.g. with well-chosen tag sequences) that facilitates its downstream applications, beginning with purification if needed and moving on to end-use in screening, functional studies or structural analysis. In biologic discovery, may be designing and expressing the drug itself, and will factor in a number of elements such as humanization of a murine mAb sequence and the amount and types of post-translational modification expected in the product.

Cell culture and fermentation scientist. Skilled at managing cell growth to achieve optimum outcomes for research or production of a biologic agent. May also assist specialized research efforts, such as growing cells on differently isotopically-substituted media for use in proteomics.

Protein purification and characterization. At the Discovery level, may require a team whose members work closely with the protein expression lab, because the key to obtaining a great protein is to express the right protein in the right host. Initially uses affinity chromatography to achieve major purification, but may have to use conventional chromatography (ion-exchange, size-exclusion, hydrophobic interaction) to remove minor impurities. Characterization is based on skills of protein chemistry, typically using mass spectrometry and the methods of analytical protein chemistry and proteomics. At the Process scale (production), purification of protein biologics will be conducted under Good Manufacturing Practice (GMP) guidelines and standards.[34]

Structural biologist. Generally uses X-ray crystallography to define protein structures, preferably with lead compounds bound to them. Collaborates closely with medicinal chemists to apply structural information to improving compound design. Also interacts closely with computational chemists. Fast-growing field of cryoelectron microscopy will also be of interest, as it gives access to structural information about proteins that cannot be crystallized.

Computational chemist. Explores the energetics of interactions between target proteins and lead compounds. Will aim to propose adjustments to lead structures with goals of achieving tighter or more selective interactions and optimizing "ligand efficiency" of active compounds. May have some favorite self-written programs, but largely uses commercially sourced platforms, such as those from Schrödinger LLC (New York, NY).

Computational biologist. Contributes expertise with modern methods of computational analysis. Usually a Ph.D. May analyze gene expression profiles and assist with design and interpretation of proteomics experiments. Also required would be investigation of GWAS (genome-wide association study) data, seeking linkage of a certain form of a gene and its protein product to a disease state, and acquaintance with pathway analysis, in which observed or hypothetical drug action is interpreted in terms of effects on canonical cellular pathways (e.g. using Ingenuity Pathway Analysis®).

Biostatistician. Both preclinically and during the clinical phase, statistical evaluation of the significance of test outcomes is a crucial activity. A 2014 editorial in *Nature*[35] and accompanying set of articles has challenged scientists to improve their awareness and performance in this area, and well-run research organizations will be attempting to do this.

Chemical biologist. Uses chemical probes and the analytical methods of proteomics to investigate drug mechanisms. Will be asked to explain the action of active compounds discovered through phenotypic screening for which the molecular mechanism of action is unknown. Strategies used may include photoaffinity labeling, activity-based protein profiling, and target affinity capture.

Metabolomics and/or fluxomics expert. Gauging shifts in the levels of a broad population of metabolites by the methods of metabolomics is, in principle, a highly promising way to evaluate shifts in disease states. Companies continue to value its potential and provide opportunities in the area. Fluxomics is the application of analytical methods to following the rates of flow through metabolic pathways. It might require partnership with an academic lab, because the effort required to set up a system could be prohibitive relative to a single discovery program.

Chapter 4: Why You Will Be Hired

Pharmacokinetics, Pharmacodynamics, and Drug Metabolism (PDM). Tracking the drug's lifetime and distribution in the body (pharmacokinetics), the timeline of its action (pharmacodynamics) and its metabolic fate (drug metabolism), either preclinically or in human patients, is a massively important activity usually organized in a department of its own. Drug metabolism frequently occurs in the liver via cytochromes P450 or glutathione S-transferases, and appreciating the action of these enzymes on a compound is a major aspect of rationalizing both its action and the duration of its action. It may also reveal why different patients respond differently to the drug. Regarding drug action (pharmacodynamics), specialized assays that illuminate what the drug is doing may have to be elaborated for each project. Principles remain the same whether the drug is a small molecule or a protein biologic, but the analytical methods employed may change. Quantitative mass spectrometry is an indispensable platform for PDM work, with the use of isotopically-substituted analytes as internal standards an everyday approach to calibration.

Formulation design scientist. Builds the final form of the medicine by combining the active ingredient with other compounds that achieve desirable properties, such as a proper rate of solubilization or release from an injection site. Factors in aspects such as particle size and crystal form, and may participate in patenting specific attributes of the product.

Analytical chemist. Looks closely at the drug substance and identifies minor components. Provides baseline for ongoing quality assurance during manufacturing. Evaluates the stability of the formulated product, a crucial property for its real-world viability. Might also defend the safety of patients and the company's intellectual property by examining counterfeit forms of the medicine. These can be identified by showing that they do not contain the characteristic mix of trace impurities found in the authentic product.

Biomarker analyst. Analytical biochemist who uses antibody-based and other assays to measure levels of a biomarker believed to signify a disease state and/or the action of the drug.

Contract work. An American company that hires you as a full-time employee commits to investing a lot more than just your salary. Normally, it will cover most of the cost of health, dental, life and accidental death and disability insurance for yourself and your dependents, as well as regularly contributing a defined amount of money to a plan that funds your future retirement benefit. It also must cover the "overhead" costs of having you around, such as parking space (maybe), an on-site cafeteria (usually), white coats, information technology (along with smart people to troubleshoot your computer problems), security personnel and the like. It's not surprising that they will expect you to work hard.

But there is a legal way for companies to avail of your skills less expensively for a limited time by bringing you in as a contract worker. In this case, your employer is an agency that links available workers to openings and then places them in the companies requiring assistance. Some of these are well known names in business, such as Kelly Services (Troy, MI) and Manpower Group (Milwaukee, WI). If you have hopes of entering a specific major biopharma company as an associate-level scientist, it may pay to find out if it works exclusively with a particular staffing agency. If it does, be sure to register with that organization.

Know what you are getting yourself into. The upside is that a contractor who makes a good impression might receive an offer of permanent employment. Entering a company as a contractor gives its management a chance to appreciate your qualities. The downside is that the earned benefits of contract work will be much less than those given to staff. You will be paid at an hourly rate, return a timecard every week, almost certainly receive no health benefits from the biopharma company (although you may be able to purchase them through your temp agency), and the legal period for which you can remain as a contractor will probably be one year at most. You will also be something of a second-class citizen in the workplace, in that you should not be "treated like an employee" when it comes to hearing confidential corporate presentations or being invited to workplace social events. Nevertheless, working as a contractor may still be a good plan, as it gets you in the door, builds your technical experience and creates contacts that can help you as you try to make the jump to staff employment. With the biopharma company having retained the right to terminate your employment at any time, you also can move on at short notice if a better opportunity surfaces.

Applying and Interviewing 5

When you apply for a job, remember that the hiring company wants you to be the perfect candidate just as much as you want the perfect job. You have your hopes, they have theirs.

If you are interested in a specific company, visit its web site and look at the section headed Careers where there will be an electronic application form. You can apply for an advertised position that catches your attention, or upload your credentials to the company's system in the hope that they will find you interesting when a position appears later.

A number of sites list opportunities from many employers in biopharma. BioSpace is an extensive site with subdivisions covering geographical "Hotbed Regions." It also links to many company profiles. SDBN (the San Diego Biotechnology Network) supplies news and job listings for the San Diego, La Jolla and Carlsbad region. If your sights are set on Boston-Cambridge biopharma, the web site of the Massachusetts Biotechnology Council has in recent years carried hundreds of job announcements at any time. Indeed.com and Monster.com are general job posting sites with particularly helpful search functions.

LinkedIn, the Microsoft-owned social media site for professionals, also offers job listings and is useful for keeping in touch with former colleagues who may be able to recommend you or provide helpful information.

In addition to entering your academic record, publications and experience, you will be asked to attach a cover letter. This lets you introduce yourself and indicate that you are a competent person.

Make the letter typo-free – have a friend scan it if necessary – and remember that people's personalities change when they go to work. Behavioral standards in the workplace are more rigid than they are outside. Avoid any hint of off-color language or references to hair-raising (even eyebrow-raising) behavior. Your pets, however charming, need not feature. Consider your social media use and delete anything that makes you look less than completely reliable. You need to be seen as somebody who will represent a company well and make its interests your interests. The best evidence that you will do this is that you have given a serious, concentrated effort in your previous endeavors, whether academic or in companies. Everything else is irrelevant.

Companies in this country must check that potential employees have the legal right to work in the United States. If your status is not obvious, do not leave them in doubt: with a folder of a hundred resumes or more, it is all too easy to move on to the next applicant. I was an immigrant to the US myself and sympathize with the difficulty, but there are so many applicants for bioscience jobs today that only truly exceptional candidates will cause a US company to petition for a change in a candidate's immigration status. It may be different in other technology fields where the candidate pool is less plentiful.

Be aware that bioscience jobs with good employers attract hundreds of applications. I have reviewed, with sympathy, many candidates' cover letters, and can offer my personal views on what goes over well and what does not. Others may advise differently. When you are an applicant, it seems best to accept reality and give the reviewing manager the information that will help him or her to an accurate decision. Bear in mind that he or she would like nothing better than to find that you are the perfect candidate. Be realistic: if a job ad specifies skills or experience that you just don't have, the hiring manager is unlikely to overlook the deficit.

Personal contacts at the hiring company are a joker card to be played with care. Consider how you judge the professional performance of your colleagues – we all do this – and remember that your friends at work have also judged you. It is entirely possible to see somebody as a pleasant and admirable person while regarding them as a walking disaster at work. If you are sure of your friend's good opinion about your own work, then consider whether their endorsement of you will be a positive or a negative. If it seems likely to be helpful, you could ask them to mention your name to the hiring manager with some positive words. But realize that nobody wants to be responsible even for causing a poor candidate to be interviewed. Having a friend inside the walls will not spare you from being appraised in the most businesslike terms.

Happy day! Your good record pays off and the hiring manager e-mails you to schedule a phone interview. The phone call will be part of a rolling process to consider candidates for personal interview, and will take 30-60 minutes. It's a get-to-know-you chat in which you need to demonstrate easy command of the science mentioned in your c.v. It is also sensible to pick up enough knowledge about the hiring company to show that you are genuinely interested in it. Then, as you will know the names of the person or persons who will call, look up their publications. This briefs you on their scientific background, and slipping in an allusion to their work conveys your interest in and respect for potential future colleagues. Finally, be ready for the hardest question of all – "do you have any questions for us?" Rather than being bamboozled, it's useful to have some innocent inquiry ready such as "how big are your lab groups?" or "do you have a regular departmental science meeting?" You might also ask the interviewers what they particularly enjoy about working at their present company, and how they think it compares with other employers.

Even happier day! Your phone interview goes well and your wit

Chapter 5: Applying and Interviewing

and wisdom leave other candidates struggling in your wake. The hiring manager e-mails an invitation for an on-site interview. If travel and a hotel are involved, the company will meet the costs and an administrator will contact you to make arrangements.

Considerate interviewers will provide a schedule for your visit at least a few days in advance. Look at the list of scientists that you are scheduled to meet and, just as you did before the phone interview, scan their publications so that you know their interests. If no schedule has appeared as the day approaches, politely request one. Unless you are visiting a deeply disorganized crew, they should already have the schedule set for their own use.

A Ph.D.-level interview that requires travel may begin with arrival on the preceding evening followed by dinner with two or three staff members. The purpose is for them to get to know you, tell you something about what it is like to work for the company, and (hopefully) put you at ease ahead of a demanding day. The senior member of the company delegation will pay for dinner – you are a guest. On the other hand, if the interview is local, the entire encounter may be limited to a single day.

Assuming you are away from home, you will spend the night at a hotel and arrive at the company premises next morning. Candidates interviewing at junior levels will probably start the interview at this point. What to wear? Business casual will probably describe the attire of the people that you meet, but a candidate usually makes the effort of dressing more formally for the occasion.

Typically you will be greeted by your potential future boss, the hiring manager. Remember that this person has spent company time and money to set up the interview and dragooned busy colleagues into accepting appointments with you during the day. At this point, therefore, you have a staunch ally. He or she is very eager to find that you are an excellent candidate, because the alternative is to have wasted people's time.

At some point in the first half of the day comes a big event – your seminar. This needs to go well and is the one occasion on which a large fraction of the group that you hope to join will hear and see you speaking about your science. As with a typical academic talk, the standard structure is 45 minutes of presentation followed by 5-10 minutes of questions. If you interest the audience as you would like, you will get questions as you speak. This is a good thing, and lets you show that you can think on your feet.

You will probably bring your own computer. Your host is responsible for being sure that audiovisual systems work properly and for getting you set up in good time so that the presentation is technically trouble-free. Especially if your slides contain animations or movies, ask in advance about the computer platform used in the company and do what you can to be sure that your slides will run as you would wish. As a Plan B against catastrophe, bring your slide pack on a flash drive with a view to loading it onto your host's computer; this can avoid the nightmare of finding that you cannot connect your own laptop to a web conferencing program or projector.

Apart from the seminar, you will march between offices to meet staff members, usually including one or two senior figures whose decision carries a lot of weight in the ultimate decision about whom to hire. Alternatively, you may be based in one spot while interviewers come and go. You can expect the tone to be pleasant and the questions to be exploratory rather than inquisitional. People working in companies need to get along with their colleagues, and this approach will extend to their interactions with you during your visit. The day will be demanding but should not be unpleasant. Typically you will meet one-on-one or one-on-two (easier on everybody) with staff members at various levels. Lunch will be taken with a group and is often a time for a more relaxed

Chapter 5: Applying and Interviewing

exchange of information about the workplace and the neighborhood. Even while you enjoy this, be sure to present your best self. If you are not used to the industrial scene, recognize quickly that the seeming informality of the workplace overlays a rather strong system of rules governing language, behavior and authority.

Throughout this day of being judged by others, you must, of course, form your own opinions about the work environment. Do people seem at ease? Do they get along with each other? Is the building clean? Is the schedule being followed properly, with interviewers showing up on time? Do people say positive or negative things about the company? Are they proud of their work and hopeful for the future of the business? Does the overall vibe feel good? If they don't offer you the job, will you still be glad to have met these people and their company?

Everything from here on depends on individual circumstances. If you don't receive an offer, at least you have gained experience that should help with your next interview.

If you do receive an offer, you will have to reflect on many things. Family should be top of that list. If the people that share your life are not happy, then you will not be happy. You will also need to consider the salary and benefits on offer and how they match up to the cost of living where the company is located. Top companies want to attract and retain the best people, and their offers will generally be fair.

Should you negotiate for better terms? This partly depends on your personality, but I would advise against it. Your request for a better offer will be considered – they want you – but money is always tight, haggling creates an adversarial start to your working relationship with the company, and all good managers want to treat every member of staff fairly. The initial offer will have been made with that in mind, and any gain that you secure by haggling may be pulled back in the process of ongoing salary reviews. Especially in today's ultra-competitive circumstances, I think it's better to get in the door and then let your talent and hard work bring you improved terms.

The case in favor of negotiating is made by Feibelman[36] in *A PhD is Not Enough!*, although he also factors in matters beyond salary such as research resources and the teaching load of a newly appointed assistant professor. He has good thoughts on these subjects.

The advice in this chapter has been aimed at the Young Scientist seeking an entry-level position, for whom responding to a job posting is a valid route to attracting an employer's attention. For more senior positions, head-hunting (the use of a professional recruiter) or personal networking becomes far more important. At the Director level and above, active recruitment of interesting candidates by a search firm intermediary will account for nearly every hire.

Joining the Team 6

You're in! You accepted the offer, you have moved if necessary, and the plan for your family is made. Time to start work.

Along with the delicate but exciting process of getting to know your new colleagues and trying to remember names and faces (harder for some of us than others), you will have many initial duties such as getting your ID card and computer and taking safety training. You will have to learn software related to essential matters such as an e-notebook and routine items like timesheets. People understand the challenges of being new and will be willing to help.

Interviewing may have shown you the local organizational structure around your job, but now you need to appreciate the broader landscape. Well-run companies worship the principle of accountability, and their pyramidal reporting structures are designed to enforce it.

For our active research worker, let's consider how a typical department is configured for the first phase of drug discovery.

Any medical area would do, but we'll imagine a big Department of Neuroscience in a major pharmaceutical company. Its function is to produce specific compounds known as "candidates" that can be presented to the company as having the potential to become marketed drugs, and it will be headed by an individual at the level of Vice President (VP). The VP supervises the work of perhaps 120 members of staff and is responsible to the company's senior scientific management for their collective performance.

No single discipline discovers a candidate drug on its own. Our

VP will marshal the efforts of both neurobiologists and chemists, and their combined endeavors will generate work for an additional team of specialists who execute complex assays to assess compound activity. Because neuroscience offers a range of distinct therapeutic opportunities, separate biology groups will be needed to deal with neurodegeneration (Alzheimer's disease and Parkinson's disease, for example) and psychiatric conditions (schizophrenia, depression). Each of these groups is headed by a Director who, typically, coordinates the efforts of 30-45 individuals. Within each group, there might be 5-7 distinct laboratories, each with its designated head and 3-6 individual contributors. It's a natural and important part of entering a group like this at any level to learn the structure, build relationships with people in other labs, and understand how the group's collective effort is intended to add up to success. A well-constructed group uses the talents and skills of individuals with a range of personal capacities, motivations and interests (Table 6-1).

If you join as an individual contributor, you can expect to feel the closest ties to your own laboratory, where your supervisor, the lab head, represents the company directly to you. There will also be a sense of identity at the group level (i.e. the unit headed by one of the Directors), in which you will normally have regular scientific meetings and be expected to speak about your own work. Meetings of the entire Department will be less frequent, and will normally be occasions on which the VP reviews the Department's progress toward major goals. They can also be necessary when an important announcement has to be transmitted through the company.

The research conducted by this group is where active drug discovery begins. Expertise in neurobiology and attention to the latest literature leads the team's biologists to hypothesize that a compound displaying activity in a particular enzymatic or cell-based assay could alter the course of an important human disease.

Table 6-1. Qualifications Needed at Each Level

POSITION	ATTRIBUTES OF GOOD CANDIDATE
VICE-PRESIDENT	Ph.D. with excellent scientific record who has been steered into a management career and gained experience at Director level. Must be willing to accept company directives and implement them as required. People skills encompass both sensitivity and toughness.
DIRECTOR	Ph.D. with excellent scientific record. Enough self-confidence to lead others through challenge without exhibiting weakness. Skilled at shaping collective effort of 20-40 people, ensuring that it is valuable and <u>seen as valuable</u> by the company at large. Sufficiently strong-willed to make difficult decisions about work or people when needed.
LAB HEAD	Ph.D. with postdoctoral experience and documented record of completed work (publications). Will be recruited because of particular skills, but also for potential future growth. Can be primarily science-driven, but understands that work of the lab team must fit into collective effort. Interprets company's requirements to lab personnel, and must have adequate people skills and personality to enable colleagues to function at their best.
INDIVIDUAL CONTRIBUTOR	Could be Ph.D., but may be at MS or BS level depending on opportunity. Recruited for needed skill set, but with expectation of future growth. Depending on career stage, needs adequate academic record or track record of successful work at another company.

They may also have the encouragement and support of a panel of outside advisors made up of thought leaders in relevant areas.

The assay group sets up a screen to search for active compounds in a large collection or "file," something that all major companies maintain and carefully curate. Advanced techniques such as high-content imaging may be applied if effects on a complex cellular function are being sought. Chemists survey the first harvest of hits from the screen, select the ones that look most tractable, and use high-speed analoging methods to prepare large numbers of close-in variants for further tests. This allows a first sketch to be made of structure-activity relationships (SAR) in the active compounds.

A swathe of chemical and physical properties of active molecules are evaluated with the goal of selecting "active matter" with drug-like physical properties, meaning that the lead compounds are likely to be well absorbed by the body when taken orally and carry no red flags that predict toxicity. If a biological agent is being sought, the process will be shaped accordingly, but the goal will remain selection of a specific candidate molecule for potential development.

Incidentally, I have encountered academic scientists (some very distinguished in their fields) who appeared to imagine that discovering a specific inhibitor for a target enzyme amounted to surpassing the main challenge in drug discovery, and that little else remained to be done. In reality, this is only a starting point, and perhaps the only problem in a typical drug discovery project that we can nearly be certain of solving. Serious tasks remaining include:

- study of the lead compound's activity in preclinical models
- determination of pharmacokinetics and pharmacodynamics (the time courses of the drug's distribution through the body and its action, respectively)

Chapter 6: Joining the Team

- identification of how the body metabolizes the drug and what activities those metabolites possess

- design by process chemists of an efficient synthetic pathway for industrial-scale production of a small-molecule drug

- selection of a specific salt form of the drug, and a particular crystalline form that gives the formulated drug acceptable properties with respect to stability and rate of solubilization

- design and optimization of any companion diagnostic test used to identify members of a patient sub-group (i.e. if it is known or expected that the drug will work in patients with a particular genetic profile, then clinical trials will be run in patients of that type to maximize the chance of demonstrating effectiveness).

All of these jobs present serious challenges and can require huge efforts of innovation by scientists. This book mainly places our Young Scientist in the early stage of drug discovery, but the later stages of the work are no less crucial. They offer great career opportunities, especially for scientists trained in analytical science or competent with mathematical models of physiological processes. Meanwhile, their scope and scale appear to escape the view of industry critics who imagine that most drugs emerge from academic science in a nearly ready-for-prime-time state. This is completely false.

The organization's senior research management must now judge whether the candidate offers sufficient potential to make further development worthwhile. The criteria will be scientific, technical and commercial. Can the selected molecule be made efficiently, so that the "cost of goods" – the cost of producing it for the market – will be acceptable? Synthetic routes devised in research chemistry labs often need to become shorter and greener to be acceptable for the long haul, and this work falls to chemists in a department devoted to inventing commercially-scaled processes. How has the market envisioned for the compound changed over the several years since work began? Are competitors ahead of us with a compound as good as our own? The commercial arm of the company will weigh in. What priority should the compound have for the limited resources available for development, especially costly clinical trials? Might the compound be co-developed with a partner so that costs and risks can be shared? Should it even be sold off to an outside company for further development because its market potential does not match the business need, or because funds for late-stage development have to go to other programs?

The individual contributor will handle a particular task that contributes to the overall effort. As a chemist, you might be completing organic syntheses: synthetic chemistry is a vital discipline in the industry, for obvious reasons. The biochemist may be conducting a sophisticated assay or, if a "phenotypic screen" with intact cells or tissues has been used, using chemical probes to attempt to understand the mechanism of action of active leads. The biologist may be conducting a screen or working with the other specialists to produce cells for relevant investigations. As a whole, the group aspires to converge on the single goal of discovering a viable lead compound, and the function of management is to unite the strands of innovation that bring that about. There should be no room for personal initiatives that disconnect from the collective effort. Alignment is everything.

If you have joined your company as a lab head, your supervisor – a Director who reports to the VP – will be responsible for delivering success in one of the contributing areas of activity. Most likely, the Director will consider you and your peers as a leadership group for the section, and expect each of you to manage your lab group in productive alignment with the Department's overall mission and goals. In a large group on the scale of our example, this will definitely involve multiple projects moving in parallel, and well-judged division of effort between projects will be essential.

Chapter 6: Joining the Team

Table 6-2. Accountability and Role by Level

POSITION	RESPONSIBILITY	REPORTS TO
VICE-PRESIDENT	Identification of specific compounds or biologicals with potential to become marketed drugs. Manages all aspects of unit's performance, budget, adherence to company policy, safety of workplace, hiring and discipline, etc. Represents company to employees as a large group.	SENIOR VP OR PRESIDENT
DIRECTOR	Coordinates scientific effort and organizational function of a group of about 30 colleagues. Maintains a leadership team composed of lab heads. Closely monitors scientific work in each lab, and ensures that the overall effort is coherent and contributes to collective goal. Represents company to employees as a team.	VICE-PRESIDENT
LAB HEAD	Intensely involved in scientific work: may or may not be hands-on in lab. Closely monitors and advises members of lab team. Reviews and approves their goals and performance. Represents company to employees as individuals.	DIRECTOR
INDIVIDUAL CONTRIBUTOR	Performs scientific work. Uses as much independence as possible, but not more. Reports and presents own work to team meetings. Responsible for proper maintenance of notebook records.	LAB HEAD

53

As a lab head, your defining responsibility is to guide and supervise the individual contributors working in the lab. Whether this position leads you to a higher level of management responsibility depends on your ambition, preferences and talent.

We all have different aptitudes, and it is uncomfortable to be asked to perform a role that does not fit our personality. Let's consider the responsibilities associated with the four levels of duty that we identified in our imagined research Department.

Just as different qualifications and talents are required at the different levels (Table 6-1), certain accountabilities are associated with them too (Table 6-2).

A Vice President accepts a high degree of personal alignment with company policy as determined by the company's most senior leaders. He or she is expected to carry out all directives handed down from higher levels. This assignment is no bed of roses. Life-changing news may have to be announced to people who are personal friends – jobs eliminated, facilities relocated, company discipline enforced when an employee has acted unwisely – and it requires strength of character and force of personality to be able to handle these duties. Nor is the Vice President just a conduit. He or she makes many decisions independently that affect the success of the Department and bear on the success of the company. Foresight, wisdom and a willingness to trust others to do their jobs well are required.

There will also be good days, we must hope. When success comes, hats are in the air and star contributors are being lionized, the Vice President must remind the team – convincingly – that each and every individual's contribution was meaningful. To be in that position is a rare privilege that should make the heavy responsibility of the position easier to carry. Companies, and everybody who works in them, need individuals with the resolve and ambition required for work in senior positions. If you feel that you are one of them and are well guided, you will have the chance to explore the territory in stepwise fashion by moving progressively through the levels.

One of the most important staging points will be the level of Director. Here you will be able to, and need to, be in personal touch with the science being done in your group. At a minimum, you will meet one-on-one regularly with the lab heads in your team, and you should be maintaining a personal relationship with every single member of your unit's staff. As this would often be around thirty people in a large company, you need a quick mind and a capacity to be interested in other people's lives as well as their work (I am told this is achievable). Apart from the personal touch, you need to be sure that your group's operations are sustainable through any reasonably foreseeable emergency. The illness or departure of one individual must not bring the show to a standstill, so is there cover for every important function required in the work? And what about team harmony? Are rivalries emerging and getting in the way of smooth cooperation?

Chapter 13 offers further comments on the gifts of personality needed to function effectively at management levels.

As a lab head, especially as you begin your career, you may combine hands-on activity in the lab with supervision of a compact team. As with all supervisory levels, it's crucial to understand that to supervise is also to serve. Being nice to people is a good thing, but to lead them to professional success and career advancement is a far greater gift. Because you represent the company directly to individual contributors, you have duties to each side of the interaction. The company is trusting you to manage its most expensive and delicate resource, its people. Those people, meanwhile, rely on you for scientific assistance in making their work productive, but even more importantly for regular readings on whether they are doing the right thing and doing it well.

Therefore, you may find it easier to ignore poor performance or unprofessional behavior in the short term, but you do your errant colleague a much greater favor if you guide them quickly back onto the right track.

Finally, if you are an individual contributor, listen closely to what your boss asks you to do, and apply your own quality control filter. When you work well in this role, you receive good ideas and make them better. Like everyone else in the company, you rely on others to be fair in recognizing your contributions, and it is important that you see how your personal effort forms part of the collective.

Lab work is not easy and there is no Undo button when the vital tube is dropped on the floor or a mental error causes an experiment to be lost. When this happened in our lab, I would point out that it was always the people who did the work that made the mistakes; the ones tapping away on their computers never dropped a tube. But, if you have this misfortune, own up to it right away and get things back on the right track.

Planning Your Work and How It Will Be Judged 7

Responsibilities assigned to team members at different levels are intended to nest into a single coordinated effort that succeeds in identifying a drug-like agent. Different projects have their own timelines, and some scientists work on more than one at any time, but the work of company staff is usually planned and reviewed on an annual basis. The 12-month cycle can begin at any time of the year, but January-to-December is convenient for discussion.

Frequent days off do not agree with scientific work, so your company may ask staff to surrender minor holidays during the year in exchange for a long, restful shutdown over the year-end holiday season. Doing this makes the New Year even more of a natural starting point for a new cycle.

A company's leaders must have a vision of its business future. For a start-up, this might be the drive toward marketing a first product or achieving progress that attracts a buyer or partner whose financial muscle can carry the work to the market. For an established company, goals for improved sales of existing products will stand alongside plans to add to them or replace them with the products of research. In addition, major companies increasingly seek early-stage potential products outside with a view to carrying them internally through full development. As the Head of our imagined Department of Neuroscience, the Vice President faces expectations of progress and productivity that contribute to this plan, and will write down goals that can be shared with everyone in the group. This allows everyone in the unit to know what they collectively aim to achieve. Besides productivity-related goals, the list will include softer items about the business culture of the unit, internal communications and scientific publication.

The work has to be divided into its component parts. Each Director follows the VP by writing down in more detail what is intended to be achieved collectively during the year by the thirty or so people in each section. This will be a more specific expansion of parts of the list specified by the Department Head. Business productivity will account for the headline issues, but there will also be cultural goals designed to orient the team for growth and any needed changes.

Next come the lab heads. They are committing their best efforts to deliver operational achievements, and need to enlist the full engagement of the individual contributors working in the labs. Goals at any level should be "achievable with difficulty," which is to say that they are not useful if they are either impossible or too easily accomplished. A business maxim suggested by George Doran in 1981 says goals should be SMART, meaning Specific, Measurable, Assignable, Realistic and Time-related.[37] This acronym is shopworn from overuse, but it is overused because it is good advice that passes the test of common sense.

Finally, each individual contributor can put on record, with the kindly guidance of the lab head, a list of goals that he or she will aim for during the year. Again, these should combine buy-in and commitment to the collective efforts of laboratory and Department with some purely personal objectives related to professional development. These could include a paper to be written or a course to be attended.

When this process concludes – each person's goals have been reviewed with their supervisor, amended if necessary, and ultimately agreed – then everyone in the Department has a set of aiming points for the year (Figure 7-1). During the goal-writing period, work has picked up from where it was before the holiday break, but setting new goals refreshes the sense of purpose that drives and steers it.

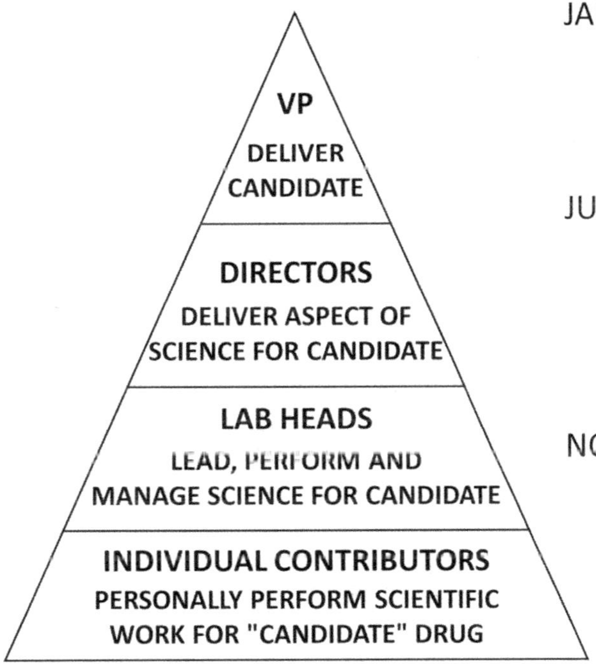
VP
DELIVER
CANDIDATE
DIRECTORS
DELIVER ASPECT OF
SCIENCE FOR CANDIDATE
LAB HEADS
LEAD, PERFORM AND
MANAGE SCIENCE FOR CANDIDATE
INDIVIDUAL CONTRIBUTORS
PERSONALLY PERFORM SCIENTIFIC
WORK FOR "CANDIDATE" DRUG

JANUARY: write goals, review, receive approval, record

- goals are aiming points, not "achieve or be fired"
- core work, plus publications, development, etc.

JUNE: mid-year review of progress – are we on track?

- ensures clear communication
- should never come to end of year and be told for the first time that fault is being found

NOVEMBER: review year's work

- self-assess, discuss with boss
- boss reviews your work for assessment by comparison with expectations and peers
- rating will influence any change in compensation

Figure 7-1. Goal setting and performance evaluation. The process provides clarity and coordination, but good sense, integrity and humanity are additional vital components of good management.

So the year moves on, work gets done, and for a few months we can work in relative peace as if time were not in short supply. But there is always urgency, not only because biopharma is an expensive business to run, and not only because one day's worth of marketing even a moderately successful drug brings a million dollars in sales, but also because patients and their families are waiting for new medicines. Reaching the end of the year with nothing accomplished would be a disaster, and so there is a need to take a formal look at progress in mid-year. Communication between lab heads and individual contributors should be fluent and frequent enough to leave no uncertainty about the progress (or lack of it) being made, but mandating a mid-year review ensures that the channels are open.

Nobody should hear for the first time at the year-end review that his or her work is below expectations: if that is the case, the problem should have been identified, discussed very clearly, and remedies identified at an earlier date. Therefore, and although it probably seems formal and bureaucratic compared to most academic labs, a mandatory mid-year review of progress can be a helpful chance for each staff member to give and receive feedback on how things are going. It can be a useful occasion to reaffirm or readjust the commitments made in the goal-setting process, and the effort required should be repaid by the benefits of honest communication.

The traditional business practice of assessing each employee's performance at year-end is not intrinsically a bad idea, but American business has had to overcome its fixation on a harsh approach to it.[38] The idea of grading all employees in a cohort against each other to form a "vitality curve," then moving to reform or fire the supposedly weaker performers became popular in the 1990's, and was widely implemented. Happily, the pendulum has swung toward awareness that formal performance reviews of any kind create so much stress that they risk doing more

Chapter 7: Planning Your Work and How It Will Be Judged

harm than good. In particular, turning employees into each other's competitors – for that is what "rank and yank" does – was not exactly conducive to building trust and teamwork. To the contrary, it encouraged people and groups to form self-preserving "silos" rather than asking how they could best promote the company's widest interest.

My own main concern with it is that it defines the desired end-state of all management as impossible to reach: that is, we manage toward a condition in which everyone is doing a great job, but evaluate our people's performance on the premise that around 15% of them are not meeting the challenge. This is consciously illogical. No expedition leader brings climbers to the base of a dangerous mountain with the assumption that 15% of them are going to perish on the slope. Why would we manage carefully selected scientists this way?

Performance review is also something of a circle-squaring exercise in that team function is emphasized through the year, at the end of which each individual is singled out for separate appraisal. The process, however mild, may be hardest on individual contributors, the people who undergo performance review but have no responsibility for the formal appraisal of others.

Professionals constantly assess each other's performance – it's as natural as breathing – and a well-informed group charged with gauging performance in a subordinate level often arrives easily at a general distribution of rankings among peers. If the company avoids the trap of insisting on unrealistically precise distinctions between performance levels ("forced ranking"), then the process can be accomplished without too much strain and with the benefit of keeping people clearly informed about their relationship to the company. But the desired end-state should be an achievable result – it should be possible for management to conclude that everyone on staff is working well.

Of course, this is not guaranteed to be the case. No biopharma company working under heavy competitive pressure will tolerate much in the way of poor performance. Management's challenge is to balance firmness with tolerance, *and to be seen to do so*, so that occasional strong and necessary actions against genuinely poor performers are not mistaken for domineering, demoralizing harshness.

Why You Might Lose Your Job

A strength of the American economy is the relative ease with which businesses rise from nothing or remodel themselves when they need to change. Economists say that this maximizes the greater good in terms of overall wealth, but the personal consequences for employees affected when a company's priorities change are painful. Coal mining and steel production are two industries in which the effects have been corrosive for entire regions of the country, but it is also not unusual to read that a biopharma company has been acquired by a rival, with jobs being eliminated to reduce costs. Even the most established biopharma companies may occasionally announce that they are discontinuing research efforts in a certain area of medicine as they save costs or switch investment to a more promising area. When this occurs, scientists conducting research in the areas to be closed are kindly thanked for their efforts and told that their jobs will cease to exist on a certain date.

After all the effort that goes into being hired and then focusing fiercely on meeting the company's goals, this is a bitter message to receive. It is probably hard to deliver as well, and some sympathy is due to those who have to make and communicate life-changing decisions to respected colleagues. After all, a successful drug that benefits many patients brings massive rewards in the marketplace. The decision to stop research in a certain area is normally made by well-informed senior managers who reckon that the work has no chance of being successful or that a rival is too far ahead to be caught.

The possibility of being "deselected" – the current euphemism for "dismissed" or "let go" – is something that our scientist/idealist should be aware of when making choices about employment. In less pressured times, major pharma companies aspired to discover medicines across all the important therapeutic areas. Today they increasingly focus on the ones in which they claim a competitive advantage and in which they view the science as adequately tractable.

The business "press," largely found online, comments daily on the activities of publicly traded companies, the ones whose stocks are bought and sold on public exchanges. A good part of this is nonsense, and you should avoid sites that are largely collections of blogs, but the journalism of serious financial papers such as *The Wall Street Journal*, *Forbes* and *Barron's* is well-informed. Relevant articles are often freely accessible on stock-quote pages. If you follow the progress of your company and its competitors, long-term trends with predictable implications are easily seen. For example, the expiration date of patent protection on a product that makes up a large fraction of a company's revenues is public knowledge. The foreseeable effects of this loss of exclusivity on sales may drive your company toward seeking to acquire a competitor or raise its own risk of being acquired, with or without the consent of its management. Either case disrupts lives, but it is usually preferable to be employed by the acquiring company.

Why does a company buy another company? A business makes a profit when it takes in more revenue than it costs to run its operations. Those costs will include (as examples) research, accounting, manufacturing, marketing and human resources departments. If the sales of the acquired company are added to the existing revenues of the acquirer, while duplicate capacity for enabling functions is eliminated, the merged company will be more profitable than the acquirer. The move can be a defensive one, in that management and the board of directors may decide to counter a downward trend in revenues by purchasing the revenue stream of another company. Cost-cutting is always part of the equation.

Chapter 8: Why You Might Lose Your Job

This system may please shareholders but creates difficulty for employees affected by the changes.

Experienced colleagues will help you build your understanding of these matters. To live entirely in the bubble of research is naïve and can leave you open to sudden shocks.

Speaking of unpleasant shocks, the scientist accustomed to the seemingly liberal academic world may take time to realize the extent to which power, authority and subordination exist in a company. The key to living with it is to be sure that your goal – mine was always to live the life of a scientist – and the company's goals are compatible. It worked for me but may not work for everyone. If you feel so strongly about choosing your own scientific problems that you cannot bear to align your efforts toward collective goals, then you may need to find a home on the academic side of the line.

The traditional source of pressure on companies is major shareholders' need to be rewarded. A newer one is the emergence of so-called activist investors. These wealthy individuals or funds acquire a significant ownership stake in a company, then use the influence that comes with that to urge management toward decisions that increase the flow of cash to shareholders, no matter what the long-term implications for the business may be. As drug discovery research is costly, risky and slow, management that appears overcommitted to it comes under criticism and can feel forced to respond by swinging the cost-cutting axe.

Management controls the day-to-day activities of the company, with a chief executive officer (CEO) at the head of the pyramid. His or her "boss" is the shareholders, represented by the company board. In some cases, but not always, the board is chaired by the CEO, making the CEO a dominant power in the company. Boards of large companies include many part-timers, persons distinguished in their own fields who bring the benefit of wide experience but are outsiders to the organizations. Boards of biotech companies often include the scientific founders and other advisors. Corporate governance is considered strong when the board is truly independent of management and contains individuals whose vision extends beyond immediate profit to include the long term health of the company and a grasp of its social mission. A strong board stands (well, sits) ready to restrain or even change management if it is not acting in shareholders' interests. In addition to profits, these can be construed as including the company's long-term social purpose of providing new medicines. The board becomes weak if management has excessive power over its composition, so that the board rubber-stamps any decision that management wishes to make.

Major decisions about exiting a research area would need to be approved by the company board under the advice of the CEO and the company's chief research officer.

Ironically, in a small company dedicated to a single project, successful development of the program to a crucial staging point can result in jobs being lost as the organization transitions to new challenges. The obvious instance would be when a medicine enters the clinic and the early-stage research that brought it to that point is viewed as complete. It would be pleasant to think that new projects will follow to occupy the workers who have done so well, but that may not be so.

The message of this chapter is that you can be doing an excellent job in every way, but still be inconvenienced by change that is completely beyond your control. Build this into your understanding of what it means to be an employee and a professional. As long as you are part of a company, you owe your employer loyalty, dedication, focus and whole-hearted effort. But if you perceive a better opportunity elsewhere, you are fully entitled in your interests, and those of your family, to accept it and

Chapter 8: Why You Might Lose Your Job

move on. Just remember that your employer retains an equal right to opt out of its connection with you.

By the way, we sometimes hear that millennial workers have little enthusiasm for the idea of long-term employment with a single organization. They are said to embrace the stimulation that comes from changing jobs every few years and to envision careers that consist of ten or more different shifts of a few years each with different employers. This may be the perspective of some of them – we are dealing with an unsupported assertion – but it remains to be seen whether the idea survives long-term exposure to reality. To me, it is a view of the future that suits the interests of employers far more than those of workers, and is a particularly shaky one when it comes to the interests of research scientists. My advice, which you are free to reject, is that long-term employment with a great company continues to have much to be said in its favor. Variety is spicy when you are thirty-three; stability tastes very good when you are fifty-five.

When you consider leaving your job to work for another company, you must take note of any "non-compete" restrictions present in your employment contract with your present employer. These are in place to prevent you from walking away with the company's confidential information and placing it in the hands of a competitor. At a junior level, you are unlikely to have much of a problem, but you may be cautioned about your obligations as you depart. In an uncomfortable case, consult your own attorney and eliminate doubt about your situation: steer clear of the legal swamp rather than attempting to swim out of it.

Finally, there are many books on how to manage other people, but few or none on the right way to accept being managed. This is strange, as if there were a thousand textbooks on how to send radio signals and none on how they should be received.

Management is the art of forming a view of the world, imagining how future actions will change it, and persuading others to make those actions reality. It takes courage for a scientist to give up personal contact with research work and make a career of leading a group to its goals though the actions of others. It is an old maxim that if you want something done properly, you should do it yourself, but being a manager doesn't follow this principle.

Being managed, on the other hand, means receiving and processing signals that come from more senior levels of your company. The ones sent by the CEO or Head of Research usually ask you to adopt certain approaches and follow recommended principles. Messages from your supervisor, on the other hand, are likely to relate to how you spend your precious time, so they need close attention and understanding.

A passive approach to being managed takes every instruction as an order and follows it as issued. We can do a lot better. It seems unfair to expect management to always be right – nobody ever achieves that – so a better approach is to play close attention, *find out what's good* about the instructions that you receive, and try to achieve the best possible outcome that combines your supervisor's ideas with your own. In this way you add value to the ideas handed to you.

The Universal Drug Researcher 9

Buffy Sainte-Marie's song *Universal Soldier*, written in 1962 and released on record in 1964,[39] casts the blame for wars onto the ordinary soldier. This is the person, in all his (male) diversity, who "really is to blame" for the horrors of war. It is a poignant poem from the heyday of the protest song, even if it is more emotionally stirring than objectively accurate.

In 2019, a war of words is raging in the news media and the corridors of power about drug prices. Prodded by news reports and opinion pieces, public representatives are summoning drug company chiefs to give testimony, and the affordability of medicines is very much in question.

Several features of the situation are causing disquiet.

One is the occasional ability of entrepreneurs to gain control of the sole source of an old, inexpensive generic drug and quickly boost its price by a massive amount. This can best be done when the commercial value of a medicine falls so low that all but a single manufacturer has left the marketplace.

Quite reasonably, this kind of profiteering is widely seen as unacceptable, and the media lose no opportunity to feed public outrage when the chance presents itself. As collateral damage, mud is slung indiscriminately at biopharma, and the entire industry has to try to try to wipe it away.

A second debate centers on the price of insulin, an essential medicine for type I diabetics and also used by many type II patients. Insulin, of course, is no longer a single medicine. Although it originally had to be extracted from animal tissues, the era of recombinant DNA-based protein expression has seen it evolve through many variants and formulations designed to improve its purity and duration of action. Despite this, it is understandable that the public is puzzled to learn that the mean price of insulin in the United States nearly tripled between 2002 and 2013 while prices for some small-molecule antidiabetic agents rose much less steeply or even fell.[40]

Insulin suppliers have answers, of course, when questioned about this situation, but the explanations tend to focus on the arcane details of US drug pricing. It is noted, for example, that (i) almost nobody purchases a drug at list price, and (ii) the great majority of patients are covered by insurance. The effect is similar to that of a smokescreen, with real clarity remaining elusive. And drug costs are only one quadrant of the mystifying labyrinth of pricing practices in US health care.

The third category of problem is different again, being the breathtaking prices that companies charge for a rapidly increasing list of novel drugs that deliver life-saving or life-extending benefits to patient groups that are small compared to the populations served by antihypertensives or statins, for example.

An important cohort is the new wave of anticancer drugs that selectively target mutant protein kinases identified as drivers of tumor growth. Good examples are Gleevec® (Novartis), Xalkori® (Pfizer) and Imbruvica® (Johnson and Johnson).

Biologicals, whether new to the market or old enough to face competition from biosimilars, can also be very expensive. Moreover, the benefits that these agents deliver in widespread diseases such as rheumatoid arthritis, breast cancer and colorectal cancer have earned them places among the world's biggest-selling medicines during the past ten years.

Enzyme replacement therapies for rare hereditary conditions such as Gaucher's Disease come at high cost, and the extraordinary new agents that improve lung function in cystic fibrosis patients are also exceedingly expensive.

Some of these medicines are described in more detail in Chapters 16 and 17.

A fourth category, also fair game for public discussion, is old drugs for which new and potent purposes emerge. Thalidomide, the cause of a medical disaster in the 1958-61 period, was later found to be effective against multiple myeloma and approved for this indication by the FDA in 2006. Patents claimed to protect its repurposing allowed it to command a new and important market. A slightly altered and improved form of the molecule, lenalidomide (brand name Revlimid®), was then introduced and became one of the world's top ten-selling medicines. Some argue that society has already paid its debt for the discovery of this class of drugs, and that fair prices for them would be lower than the prices charged.

A free lunch is always pleasant if the chef is on form, and the next best thing to a free lunch is an inexpensive one. When it comes to health care – a central component of a good life in economically advanced societies – everyone instinctively sympathizes with the individual or family affected by illness and would like their needs to be met irrespective of cost. The same sentiment extends to persons in developing nations. Those of us who live in the relative economic comfort of the US or Europe often experience shame at the realization that large parts of humanity lack services that we consider indispensable.

Therefore, the cry goes up that Something Must Be Done. But it nearly always should be done by Somebody Else, and the biopharma companies are variously criticized for pricing their medicines beyond the means of developing countries and failing to conduct adequate research against diseases that predominantly affect them. The implication, sometimes but not always thought through, is that biopharma companies should reduce their costs and profits in order to meet the human need that clearly exists for their products.

Even in developed countries, and especially between companies and national health systems beyond the US, drug pricing generates conflict. Again, the natural public response is to castigate companies for excessive pricing and profiteering based on human need.

In due course, these arguments will have to be resolved in the arenas of politics and business. We cannot settle them here. But what is the researcher to make of them?

The Universal Researcher wants to dedicate him or herself to finding new and better cures for disease so that medicine can move ahead. He or she can't do that without a salary and the tools and conditions required to do the work. Drug discovery is a demanding, long-term exercise that requires total focus. What role will you, as the hands-on researcher, have in the debate about pricing?

Here we need to sip the ice-water of reality. Biopharma companies are not in the least interested in seeing or hearing their employees make independent public pronouncements about the rights and wrongs of their business practices, including pricing. If you enter the industry, you will need to accept that working for a company involves accepting its rules and disciplines. A central expectation is that you will not vent opinions in the public square that damage the company in any way or complicate the message that its leaders are attempting to put forth.

Does this leave the researcher (metaphorically) bound and gagged, with his or her integrity sold for the sake of a salary? Fortunately, that is not the case at all.

Chapter 9: The Universal Drug Researcher

To the contrary, companies usually beg their employees to tell the truth about drug discovery, which means educating their families, friends and neighbors about where medicines come from and how massively demanding is the process of discovering them.[41] People who know you and see how hard you work will generally understand that you are not frittering away your time. It will also cross their minds that employers are not in the habit of keeping people on a payroll for the sake of good appearances. Just by doing your job to the best of your ability and speaking about it honestly, you will help to spread the truth about drug discovery in the industry. In the end, wider appreciation of the truth should lead to the best outcomes for everyone.

I am not calling for companies to be left alone to do as they please. As we said earlier, companies are driven to be profitable just as dogs are driven to bark, and there will always be an instinct to maximize that profit. What we all need is for that instinct to be moderated by leaders who understand how to combine it with a sane sense of sustainability and an appreciation that companies – contrary to business thinking that has become too prominent in recent years – have responsibilities that go beyond maximizing the return of wealth to their shareholders. In such a world, the Universal Researcher will be able to give society the gift of his or her dedication to the advance of medicine and human welfare.

Finally, and with reference to the first problem named – the piratical inflation of the price of special-market generic drugs – an innovative response has been the recent foundation of a not-for-profit generic drug company that will attempt to meet part of the need while maintaining affordability for patients. Civica Rx is to be based in the State of Utah and sustained by a coalition of health organizations that includes an extensive subset of the nation's leading hospitals.[42] The additional engagement of philanthropic foundations suggests that the company will not be intended to fund itself entirely by product sales. Moreover, its leadership states that it does not hope or intend to drive established generic makers out of the market, but rather to promote price stability by ensuring that there will be at least one supplier in the market for which profit is not the principal aim.

The possibility of producing drugs on a not-for-profit basis reflects the accumulation of modern medical knowledge over a long enough period that the innovators of these medicines should long since have reaped their fair reward. By all appearances, the motivation behind this initiative is honorable and its intended effects should be positive.

It should be clear, as a counterpoint, that performing *innovative* drug discovery on a not-for-profit basis would be an exceedingly difficult undertaking. This can be done on an occasional targeted basis for important global diseases, as we will note in Chapter 12, but requires the charitable support of a major sponsor.

Presentations and Publications 10

Basic research is the feedstock for drug discovery, and biopharma companies make it a priority to give their scientists broad access to journals. But, in addition to being voracious consumers of the scientific literature, they are also significant contributors. Allowing company researchers to publish their results has important benefits. The company boosts its scientific reputation, and shows scientists considering employment with it that signing up is not an express ticket to anonymity. Meanwhile, the publishing scientists gain career satisfaction and build external reputations as respectable contributors.

Another benefit is to burnish the image of the company's research *within* the company. Business leadership is often commercially rather than scientifically trained, and it helps when a research chief can show that its internal research merits publication in top journals.

When a compound advances into clinical development, publication concerning it acquires a different purpose and character. Papers from this stage of the process document the medicine's performance in peer-reviewed medical forums, validating its future clinical value. These papers are prepared and managed with extreme care, quite unlike the scientifically rigorous but otherwise relatively unrestricted reports authored by early-stage researchers.

Disclosing the company's early-stage research must not damage its claims to intellectual property. Because novelty is crucial to patentability, companies must carefully control the disclosure of their research results to the outside world. There will normally be a process in which the lead author submits for internal review a manuscript or slide deck that he or she proposes to disclose, and management evaluates effects on ongoing patent applications and other competitive factors. In some cases, even disclosing a company's active interest in a target might be considered unwise, and clearance to publish could be withheld for a time.

An especially grim prospect for biopharma companies is any misadventure into scientific fraud. Imagine the reputational damage that a company marketing a cancer drug would suffer if one its researchers was shown to have fabricated relevant data. FDA and competitors would be quick to take notice. This factor reinforces the need for careful management of scientific publications.

Even so, researchers entering industry should feel positive about their prospects of being able to publish. Companies' deep interest in their own reputations and their need to keep attracting talented researchers has kept publication from industry very much alive. You can verify this by submitting the name of any major biopharma company to a resource such as Google Scholar.

It is one thing to receive clearance to speak or publish, another to do it well.

Sir Peter Medawar wrote amusingly about the sleep-inducing power of scientific talks, but he never foresaw the ultimate weapon – PowerPoint. The slides projected during his own talks would have been made by sandwiching a small photograph of a printed or handmade graphic between glass windows. These took some making, but did the job: in the 1980's, I heard the great chemist Linus Pauling reminisce about discovering the protein α-helix some thirty years earlier. He showed his original slides, one with a pronounced crack running through the glass.

We still speak of "slides," but only as a nod to the past. Now that PowerPoint rules, the graphics used in scientific talks include vivid colors, animations and movies. Audiences may still nod off – scientists work hard – but the new technology certainly makes it easier to prepare a talk. Even so, most younger scientists find it challenging to present a seminar.

Researchers are no meaner than any other professional group, but they come to your seminar in a skeptical frame of mind, ready to advance the march of science by showing you the flaws in your argument. Friends who recently asked kindly about your children and pets may turn savage if they consider your experiments inadequately controlled. Expect nothing else!

To some extent, scientific presentations are a contest between the speaker and the audience, with direct questioning as the battleground. This is the scientific way, and it needs to operate in a company group just as much as it does in a university. Despite this, it is unpardonable for a senior scientist to destroy the confidence or seriously injure the feelings of a junior colleague. The need for frank scientific discussion never excuses bullying, and a colleague's level of experience must always be taken into account.

We can all admire the fluency of a speaker who warbles through fifty minutes of complex material without hesitation. Most of us have also endured the horrible experience of being in the opposite case, where we stumble through a talk to the point at which we are ready to apologize to the audience. What separates the two types of speaker? A wide vocabulary and a talent for felicitous expression are helpful, but the real key to being comfortable on your feet is to *know your material*.

There are things in life that you find it easy to learn about. It might be music or fashion or a favorite sports team. If you had to brief an audience about it, and had pictures to help, you would probably manage pretty well. The challenge in science is to command the subject material so well that fluency is natural. My advice is to be patient with yourself and not worry. You probably have perfectly adequate fluency and vocabulary, and any early difficulties stem from the complexity of the science. The point of your training is to master this. Give yourself time, and you will grow more comfortable before an audience.

Having said that, it is always possible to get a fair question that you cannot answer. In that case, it can work well to say something along the lines of "that is a good suggestion, but I don't think anyone has tried the experiment; I will definitely look into it when I get back to the lab."

Many capable scientists deal with the extra challenge of communicating at work in a language that is not their native one, and that they may not even speak at home. I am a native English speaker and not the best person to advise them on this challenge, but I appreciate its scale when I recall my total ignorance of Asian languages and my limited competence even with French or German. Fear of causing offence may limit discussion of this issue, but it deserves more attention. Some interesting resources do exist to help scientists who view it as a barrier, such as a coaching company in the Research Triangle area (Triangle Speech Services) and CD-based programming (Pronouncing American English), but I am not in a position to endorse or comment on these services or products.

PowerPoint is a fantastic tool, but needs to be used with care. Animations that look clever on your desktop can slow your presentation in the conference room. Save them for when they really add value. Beware of traps such as inserting images by hyperlink; if you are not on the internet when you give the talk, they will not appear. Use clip art sparingly or not at all. Be a rebel and abandon templates with space-hogging company branding; the intellectual giants listening to you will remember where you work without being reminded by every slide. And, most important, don't overload the slides with your own content; your listeners' thoughts stray easily to their own work, and strong, clear messages about yours are needed to hold their attention.

Slides give you a route map for the presentation. If your points are well-ordered and the supporting evidence adequate, you will not flounder. Practice until you have found the right words for each item, and plant prompts in the slides so that those words come back to you. Relax as best you can, and then lead the audience through your story. Complicated experiments can be mapped out with flowcharts. If you are in an academic group, practice with your lab mates or ask your supervisor if you can rehearse at the group's next meeting. Humor is allowed if you can carry it off, but your listeners have come to be educated rather than entertained. If you have done your thinking before your talking, it will be fine. You will deserve a round of applause and the murmurs of "well done" and "nice talk" that follow from your colleagues.

As for content, you do not need to present an exhaustive introduction to your project. The audience does not need to be informed that your boss is a titan of the field: they care about *your* effort. Set the scene in a few minutes, let them know what part of the problem you set out to solve, and then tell them how you went about it.

This advice is all very well, by the way, but it cannot compensate for a weak package of work on a topic in which even charitable listeners fail to see significance. If you are looking ahead several years to the time at which you will give career-critical presentations, assure yourself that you are not drifting into a scientific dead zone. Observe the career prospects of people a few years before you in your group, and make the hard decisions now that will give you a chance at your dreams.

A talk is provisional, but a scientific paper is written to last forever. That makes it a serious matter, and asides that can lighten a seminar and keep an audience awake seldom fit in a published paper. A paper should also be your final statement on its subject, and errors are less excusable.

Apart from the credit that it should bring, publication has an ethical component. Company-based research consumes a lot of funding, even if it comes from private investors, and it is a fundamental value of science that most research should not need to be done twice. Carefully recording your work and sharing it allows science to move on.

The main thing to remember is that scientists from biopharma publish a lot of work, and you will be able to join them if you can muster the willpower to make the considerable effort required.

Numerous books are available on how to write a paper, so I will give my own recipe very briefly. For me, the process tends to be long and arduous with endless revisions, but it begins with a document framed by the usual major section headings of Title, Abstract, Introduction, Materials and Methods, Results and Discussion, Acknowledgments and References. I start by drafting a working title and writing an introduction, and I might plug in accounts of key methods early on to give the newborn manuscript some encouraging volume. But the crucial work is to assemble the figures and tables that provide the foundation of your Results and supply the basis of the story that the paper is to tell.

Scientists are usually on their own when it comes to making figures for their manuscripts, even if the largest companies still maintain in-house graphics studios. Many journals specify that figures must be in a graphic format, often .TIF (Tagged Image File Format) at a resolution of at least 300 dpi (dots per inch). If you compose figures using Microsoft PowerPoint, you can use the Save As function to convert one or more slides to .TIF format, but the standard resolution of the output is only 96 dpi. At this book's

Chapter 10: Presentations and Publications

time of writing (2019), you can change this to 300 dpi following instructions provided online by Microsoft (try searching for "PowerPoint output resolution"). Be very careful when you do this or request help from your Information Technology department. Once the change has been made, you will create .TIF files with the needed resolution, and you can crop white space from their edges using the Photos app in Windows 10.

More sophisticated graphics may require use of more advanced resources than PowerPoint, such as a combination of Adobe Illustrator and Adobe Photoshop.

Consider the final size of your figures in the journal, and print them at this size for review as you build them. Journals usually give guidelines about font size and the like in their online Instructions to Authors, and these should be followed.

Some other key tools and skills come into the writing process. Reference manager software such as EndNote (Thomson Reuters) is virtually indispensable, and the effort involved in learning how to use it is thoroughly worthwhile. Its two main functions are to collect references from the literature into libraries stored on your computer and to place properly formatted citations and references in your manuscript. The intake function is enabled by all major journal web sites, so you should never have to type a reference into the database. The power and versatility of the output function causes some complexity, but it can be fun to master the skill of getting your references into the right format for your journal of choice. Then, in the not unknown case that your paper is rejected by the first journal that you try, a few clicks allow you to change the format of citations and references to be ready for the second one.

Speaking of rejection, I liked the thinking of a colleague of mine who pointed out that it makes sense to start by aiming for the most prestigious journal that you can reasonably imagine accepting your paper. If immediate rejection without review follows, then you may have been overoptimistic, but not much time will have been lost. If your manuscript is thoughtfully reviewed, you may be able to improve it or have the reviews considered by the editor whose journal you try next.

There are fields in which you can anticipate the identity of at least one likely reviewer of your manuscript. It will do no harm, and may do much good, to reference the contributions of that august figure in your manuscript. Vanity is no stranger to the scientific world, and insecurity is not always put to flight by success.

Finally, proper selection of authors is important and occasionally causes discomfort or even a dispute. In principle, there should be no difference between academic and company science when it comes to naming authors. In practice, company groups may be more open to including coauthors who made purely technical contributions that were indispensable to the project, even if they might struggle to display full command of the scientific issues. The most widely accepted set of guidelines on this issue is given by the International Committee of Medical Journal Editors (ICMJE),[43] and also extends to matters such as financial disclosures and conflicts of interest. To exclude a contributor who should have been included appears to be a worse error than including one whose contributions were minor, but that rationale is not to provide cover for gratuitously including a senior figure (for example) who hovered high over the work without contributing to its specifics.

Lifestyles of the Big and Little Companies   11

Starting out small and making it big is one of the great American dreams. Technology companies – Apple, Microsoft, Dell, Google, Facebook and the rest – are obvious examples from recent decades, but biotechnology has had its own spectacular ascensions. Genentech (now part of Roche), Amgen, Biogen and Regeneron are examples. There is an obvious attraction to the idea of starting in ragged blue jeans, or their business casual equivalent, and growing a company to the point at which the only uncertainty is the color of your new Ferrari. Founders and low-number employees with significant stock positions are incentivized by this dream, but they need a tolerance for risk.

As companies rise, others may fall. When the new tech giants grew exponentially, they sent an earlier generation to the boneyard. Once-great companies such as Digital Equipment Corporation, Wang Labs, Data General, Hewlett-Packard and IBM were acquired, split or forced to reinvent themselves under the onslaught of innovation. Marveling at today's biopharma gold rush in Greater Boston, we should recall that the same region thirty years ago was comparable to Silicon Valley as the birthing ward for new computer-based industry.

In biopharma, the story is less about displacement of existing pharma companies and more about the emergence of biotech as a second channel of innovation, distinct at first and later entering a single mainstream. Yet again, Nobel Prize-winning basic research set the stage for rapid action in industry. This was the discovery of restriction enzymes and ligases which, assisted by Sanger-method DNA sequencing, allowed cloned DNA to be cut and joined up to order. This initiated the field of genetic engineering and spurred the emergence of start-ups that saw the potential for a new business. In a textbook demonstration of how the ambition of incentivized inventors can lead to indisputable public benefit, the biotech industry rose from small shoots in the late 1970's to flourishing strength at the turn of the century. First, it created protein drugs that replaced factors that the body lacked or were administered in medical emergency. These included insulin, human growth hormone, erythropoietin and tissue-type plasminogen activator. Later came interceptors of disease-driving signaling proteins, both soluble receptors and monoclonal antibodies, and agents that are themselves signaling factors.

Some traditional pharmaceutical companies kept their distance at first. They were entering a golden era in which small-molecule drugs – costly to discover, but often less expensive than biologicals to make and market – would revolutionize the treatment of widespread conditions such as elevated LDL-cholesterol levels (statins), depression (selective serotonin reuptake inhibitors), and heartburn (proton pump inhibitors). With the scope of the established side of the business expanding, it may have seemed to some companies that biotech drugs could remain in a biomedical minor league.

From this perspective, recombinant DNA-based methods were more interesting as enabling technologies that could boost the discovery of new small-molecule drugs. Their availability for this purpose was crucial to the industry's effective response to the emergence of the AIDS epidemic in the early 1980's. The development of HIV-1 protease inhibitors, which in western countries first transformed the infection from a death sentence to a chronic but survivable condition, depended heavily on structural biology powered by recombinant DNA-based methods.

Eli Lilly (Indianapolis, IN) had good reason to take a different view. As the leading vendor of insulin sourced from animal tissue,

Lilly had a direct interest in Genentech's ability to produce human insulin by recombinant expression in bacteria. This led to a business agreement between the two companies that gave Genentech a crucial boost toward survival and profitability in its early years.[44]

The evident success of biological drugs changed minds in due course, and some (not all) prominent biotechs were absorbed by traditional pharma companies that envisioned a future based equally on large- and small-molecule drugs. Boston-based Genetics Institute was taken in by Wyeth in 1996, and Wyeth became part of Pfizer in 2009. ImClone Systems was absorbed by Lilly in 2008. Genentech became part of Roche in two steps, with 60% acquired by the Swiss company in 1990 and an option to buy the remaining 40% exercised in 2009.

Even as the fortunes of large companies have risen and fallen, the notion of turning leading-edge biotechnology into medical advances and profits has endured, although investment levels fluctuate with the financial climate. Should our aspiring Young Scientist looking for career opportunity be drawn toward large, established companies or seek a hot opportunity at a small company that might be the Next Big Thing?

It depends on what you want and need from your career. The obvious guess is that big companies combine the best job security with a tendency to limit your opportunities, whereas a senior position in a start-up company balances precarious employment against a small chance of becoming wealthy, and even a little famous. There is some truth in this, but people who crave security (no crime) must realize that the level of risk has risen far above zero in even the most established companies. As we saw in Chapter 8, management decisions beyond your control may vaporize your job, no matter how well you have been doing it. On the other hand, a small company that bets everything on one idea, perhaps even on one project, has an obvious risk of crashing and burning with what NASA calls loss of mission, vehicle and (with respect to employment) crew.

The choice comes down to personality, opportunity and preference. Big companies in biopharma virtually guarantee nice working conditions, human resource departments that strive to comply with company policies and employment laws, positive intentions to achieve and nurture a diverse workforce, steady budgets for scientific work based on predictable revenue streams, and good practices related to laboratory safety. On the downside, you may also be on the receiving end of corporate indoctrination as people in senior levels of the company download the genius of their high-priced consultants to rank-and-file employees. It is part of your job to put up with this stuff pleasantly, identify the good parts, and apply them.

Small companies may accommodate more spontaneity and improvisation, but this almost inevitably coexists with financial risks. If a key investor or venture capital fund loses faith, the company runs out of fiscal fuel and drops you back into the job market. In very small companies, there can also be the issue of dominance by or dependence on one individual, neither of which promotes stability.

Get a sense of how finances run in large and small companies. In a traditional large pharma company, compensation varies with the level of responsibility, and may come as a combination of salary, bonus and stock-related benefits that are deferred to encourage employee retention. Other benefits such as health insurance can also be expected for staff, but contract researchers usually will not receive these.

A small company begins with a combination of scientific inspiration supplied by one or a few researchers and seed money that they contribute themselves or obtain from an outside "angel"

Chapter 11: Lifestyles of the Big and Little Companies

investor. Ownership of the company needs to be divided up, and it is the basis of entrepreneurship to plan that part of the company will eventually be sold to additional investors, with the proceeds enriching (to some extent) the founders. The ultimate step in this regard is to "take the company public," which means that a large fraction of its ownership is auctioned off to the public through a stock market such as NASDAQ. This is the initial public offering mentioned in Chapter 2. Doing this successfully requires that the new investors believe that they are receiving good value for their money as judged in terms of future business success.

For the small group of company founders, this trajectory can produce a major financial reward. Employees who have joined as scientists may, if they are early enough or important enough, have received stock in the company that allows them to enjoy a share of the benefits. But if they are rank-and-file employees who provide steady contributions at the bench but appear to be replaceable, no special reward beyond continued employment may stem from the company's progress. If you dream of becoming a biotech multimillionaire, you need to start your own company or become an early, high-valued employee who is crucial to a company's success.

Academic Drug Discovery 12

The existence of a vast and profitable biopharma industry is not a morally essential imperative without which society would collapse. But if we accept that the economy's invisible hand creates the most efficient available mechanisms, we deduce that the industry exists because it is *necessary*. If there was a better way for us to develop new medicines, it would send biopharma as we know it into retirement, just as old-guard computer companies were swept aside in the 1980's. For example, if university and government scientists could manage the task in a not-for-profit setting, drugs *might* be much less expensive than in today's free-enterprise market.

Dissatisfaction with the current paradigm has persuaded leaders in some academic institutions and the NIH to challenge it by starting up their own discovery operations. This is a brave venture and, if nothing else, provides a space in which drug discovery can be attempted in medical areas that companies find scientifically intractable or that offer insufficient economic incentive. An example of the first might be complex neurological diseases. Two major examples of the second would be antibiotics and drugs for diseases that mainly occur in tropical regions.

The Drug Discovery Laboratory at the University of California, Los Angeles, seeks a therapeutic approach to Alzheimer's disease (AD) "based on the hypothesis that AD results from an imbalance in an extensive array of networks."[45] It advances small-molecule therapeutics that can either be entirely new, and produced by in-house chemists, or be previously approved medicines with potential to be repurposed. When applicable, of course, the second strategy offers the huge advantage that the compound's safety

Chapter 12: Academic Drug Discovery

profile is known. It also bypasses the typical preference of pharmaceutical companies for drugs that are entirely new and able to be surrounded by a secure fence of intellectual property rights.

Another liberating aspect of this approach is to avoid being locked into biopharma's preferred strategy of monotherapy, the use of a single agent to modulate (at least in theory) a single target by a single mechanism. If a combination of two old drugs happened to provide benefit in preventing or treating AD, just as low-dose aspirin has been thought to prevent heart attacks in certain patient groups, this immeasurably valuable discovery would be hard to make into a commercial blockbuster. It is also unlikely ever to be discovered by a major pharma company. Given the troubling difficulty encountered so far in drug discovery for AD, the willingness of university researchers to explore the territory has to be welcomed. Biopharma may do what it does well, but it does not do everything. Academic work that explores related but distinct approaches strengthens the overall effort.

An NIH-based program called the Accelerating Medicines Partnership (AMP) began in 2014 with five-year projects in four disease areas (AD, type 2 diabetes, rheumatoid arthritis and systemic lupus erythematosus.) A project in Parkinson's Disease started in 2018. Twelve biopharma and thirteen nonprofit organizations joined FDA and NIH in this initiative, much of which is directed toward identifying targets and biomarkers using the emerging tools of genomics and related disciplines. This is a precompetitive activity from the point of view of companies, but the possibility exists that a target for which active lead compounds already exist in another program will turn out to be relevant to a disease surveyed by AMP. These compounds will then be repurposed into that area.

AMP is sponsored by NIH's National Center for Advancing Translational Sciences (NCATS), a multipronged initiative that includes drug repurposing as one of its efforts to "bring more treatments to more patients more quickly." Some experienced and knowledgeable leaders of biopharma have wished AMP success, but emphatically stated that they regard drug repurposing as unpromising.[46] NCATS, from this perspective, is seen as trying to improve on activities to which biopharma already applies vast talent and resources, with negative impact on support for NIH's indispensable basic research function (including the funding of grants).

At the Simmons Comprehensive Cancer Center at the University of Texas Southwestern Medical Center (Dallas, TX), the use of high-throughput screening (HTS) in as many as 250 projects has led to licensing for clinical development of at least six candidates or sets of leads. HTS was the central tool used by pharma companies for lead discovery from 1990-2010, and is applied at Dallas in tight connection with evaluations of other aspects of lead molecules, such as those related to ADME (absorption, drug metabolism and excretion). It can only be a positive that niche opportunities for drug discovery are able to be explored through the operation of this facility, which to some extent emulates the operations of a small biotech company focused on early-stage discovery.

At Johns Hopkins University (Baltimore, MD), the Johns Hopkins Drug Discovery Program started in 2009 with a focus on neuroscience and was expanded in 2015 to operate in "a wide range of human disorders including drug discovery projects in oncology, immunology, neurology, psychiatry, ophthalmology, and gastrointestinal disorders." The program started by recruiting experts from biopharma to bring in capabilities related to drug discovery, and in 2018 spun off the start-up company Lorem Therapeutics to advance drug discovery in the cancer area.

The Department of Pharmacology at Yale University (New Haven,

CT) deploys a formidable range of assets for drug discovery. It uses target-based screening approaches with small-molecule compound libraries, and also takes input from "structural biology, integrative cell signaling, and neuroscience programs." Large-molecule approaches are also explored when suited to the target.

At Harvard University Medical School (Boston, MA), the Harvard Program in Therapeutic Science (HiTS) begins with the view that traditional drug discovery approaches are too much invested in reductionism – that they oversimplify the problem by adhering to the "magic bullet" concept under which a single agent hits a single target. The alternative being explored at HiTS relies on the emerging science of systems biology, in which the metabolic or informational signaling traffic through multiple related pathways is assessed quantitatively so that side effects and polypharmacology can be factored into projections and rationalizations of drug action. This program, therefore, is attempting to reinvent the process of drug discovery, a highly appropriate mission for an academic center located in one of the world's great biopharma clusters.

The web site of the Academic Drug Discovery Consortium (ADDC) lists 150 centers or programs interested in drug discovery.[47] For the Young Scientist whose career goal is to enter biopharma, one way to experience advanced ideas that could be part of biopharma's future might be to spend some time at one of these centers. It would be very important to be sure that you are joining a group with genuinely advanced ideas and high academic standards rather than an operation that weakly imitates what has been done in industry for many years.

The lapse in the discovery of new antibiotics is a serious problem that needs attention immediately and is not getting enough.[48] Few major pharma companies persist with work in the area because drugs that are only taken by the patient for a short time have poor prospects for profitability. In addition, new antibiotics now need to be held in reserve to treat patients infected with microorganisms that resist older agents, further restraining their market penetration. The alarm has been raised, and action is needed to restore the incentive for new discovery.

Occasional reasons to cheer have come from reports of possible new classes of antibiotics discovered by academic groups.

A 2015 report from an international collaboration that included a group from Northeastern University (Boston, MA) described how a new strategy for growing bacteria that live in soil, but could not readily be cultured in vitro, led to the discovery of teixobactin. This peptide-like antibiotic disrupts the growth of bacterial cell walls by binding to one of their lipid precursors.[49] Antibiotics that target bacterial enzymes or ribosomal RNA are prone to the mutational runaround that leads to resistance, but teixobactin's lipid-targeting mechanism seems to make it much less susceptible to this problem. The compound was licensed to NovoBiotic Pharmaceuticals (Cambridge, MA). Simplified analogs have been synthesized and showed promise when tested against an eye infection caused by *Staphylococcus aureus* in mice.[50] Preclinical development continues at the time of writing in 2019.

Another strategy for defeating resistance is to introduce drugs that block its mechanism, and thereby resensitize bacteria to the original antibiotic. A consortium of labs at Washington University School of Medicine (St. Louis, MO), has reported structure-assisted discovery of an inhibitor of a flavin-dependent redox enzyme dubbed tetracycline destructase because of its role in producing resistance.[51] This strategy for defeating resistance is not new: β-lactamase inhibitors have been marketed for several decades to block the actions of hydrolytic enzymes that created resistance to penicillin-like antibiotics and others that feature a reactive four-membered β-lactam ring.[52]

Drug discovery in biopharma is driven by the twin incentives of medical advance and business success, which have been linked with a platinum chain. In the academic world, motivation might be more variable. Career progress depends on scientific quality and timely advances past checkpoints such as degrees and tenure, but the golden carrot at the end of the rainbow is the opportunity to make one's academic research the foundation of a biotech company. Many universities encourage and facilitate this process, as the rewards for success are plentiful enough to satisfy all parties. The progressive attitude of the Massachusetts Institute of Technology to this process, first in military and electronic technologies and later in biotechnology, undoubtedly fueled the prodigious success in both fields of its surrounding region.

So, is the stumbling, lumbering *Pharmasaurus* on its way to extinction, soon to be outmaneuvered by nimble, warm-blooded biotechs? Certainly not in a single cataclysm, but important change is certainly afoot.

We are about to see – we are already seeing – a shift in the structure of early-stage drug discovery. Traditionally a treasured in-house activity of biopharma, early discovery will increasingly take place in smaller organizations. The results will be acquired opportunistically by bigger companies when something promising is ready for development. Already, some large pharma companies have announced cutbacks in their early-stage research in difficult areas and shifted the savings to venture capital units that evaluate and fund promising research originated elsewhere. This is not the greatest news for Young Scientists, but the business logic of the approach is undeniable. So do not deny it: see if you can work with it. Given the intrinsic – indeed, essential – fragility of smaller companies, it may become a fine art to make the best judgment about which ones to consider joining, and in what locations.

A world in which early-stage drug research largely happens in small companies necessarily causes the relevant people and skills (which are not limited to science) to accumulate in clusters. This allows researchers to change their employer every few years without having to relocate. But note the challenge for the Young Scientist and his or her slightly Older Self: every enforced venture into the job market carries the hazard of being compared to the next wave of talent, and sustaining a full career will take a lifelong effort of continuous learning and self-awareness.

Established biopharma is frequently accused of concentrating its efforts on medical issues of the developed world, where sufficient wealth exists to fund profitable markets. This emphasis has been real, and can be understood as a response by companies to the relentless demand for profits and efficiency from their investors. Change is in the air, however, and a positive (although fragile) trend toward worldwide access to medicines is described in a 2019 report by the Netherlands-based nonprofit Access to Medicine Foundation.[53] Interestingly, examples of progress come when companies understand that genuine business opportunities exist in serving patients in less developed countries: this is not a question of asking companies to turn themselves into charities. By working with governments and well-informed international expert groups, companies can advance public health in a way that is entirely aligned with their proper purpose. Everyone knows of the need for more successful treatments of tropical communicable diseases, but an important part of the newly recognized opportunity is to deal with noncommunicable diseases such as diabetes and cancer.

Regarding charitable efforts, there are too many examples of international outreach for any list given here to be complete. For example, drug donation initiatives by Merck and Co. and by Pfizer Inc. have each supplied many hundreds of millions of doses of anthelmintic[54] (Merck, for river blindness) or antibiotic[55] (Pfizer, for trachoma) agents used to prevent or treat two different major causes of blindness in African and Asian countries.[56] More direct

Chapter 12: Academic Drug Discovery

efforts at advancing human health include Merck's Ebola vaccine, rushed to the scene of an outbreak in the Democratic Republic of the Congo in 2018, GSK's ongoing efforts in tropical disease research at its Tres Cantos research center in Spain, and Novartis's contributions to developing antimalarial medicines.

Despite these fine efforts against communicable diseases, in which the companies have worked closely with the World Health Organization and national medical authorities, it remains true that the western-model business-driven biopharma industry does not, and cannot be expected to, devote more than a fraction of its energy to neglected diseases of the global South.

This is widely understood, and outstanding work has taken place in academic centers to address the obvious need. An especially creative one has been a sustained assault on Chagas disease over many years at the University of California, San Francisco (UCSF), where it benefits from running alongside world-leading research in the fundamentals of protein-ligand association. Chagas disease is a serious parasitic infection with the protozoan *Trypanosoma cruzi* that infects as many as 8 million people in Mexico, Central America, and South America, and is also found in Caribbean countries.[57]

Finally, another model is represented by the Bill & Melinda Gates Medical Research Institute (Gates MRI), self-described as a not-for-profit biotech company that will be located in Cambridge, MA, and Seattle, WA. According to its website, the mission of Gates MRI as of 2019 is "to eradicate malaria, accelerate the end of the tuberculosis epidemic, and end diarrheal deaths in children." Initial funding provided by the Gates Foundation will ensure that this initiative can set its course according to global medical needs rather than market opportunity.

Should I Stay or Should I Go? Moving Into Management 13

Some people want to keep on doing exactly what they are doing. There's nothing wrong with that: if you have found your dream job, you should keep on doing it. Others feel ambition to rise to a senior position, which requires climbing a ladder of increasing responsibility. There's absolutely nothing wrong with this either. It's a question of what is right for yourself and for the people you work with.

Companies of any size have defined ladders of jobs, each with a set salary range and accompanying benefits. If you join a company and do well for a couple of years, you may receive a promotion designed to retain your services by increasing your compensation and persuading you that you are in the right place. This is a logical move for your employer, because replacing a competent, experienced employee is expensive and disruptive. That particular rocket has only one or two booster stages, and it burns out quite soon. It will take you to a certain orbit that you can expect to maintain for the rest of your time with the company. If that suits you, and it suits many people who require or prefer a certain work-life balance, no problem.

This apart, companies no longer promote staff just because they are doing a good job. Promotions are driven by business need, meaning that they correspond to an employee accepting increased responsibility or clearly working above the level of their present position. Perhaps a departing or retiring manager needs to be replaced, or a new research initiative needs an accountable individual in charge.

Chapter 13: Should I Stay or Should I Go? Moving Into Management

Persons aspiring to the Director or Vice President levels usually must move between companies to achieve the necessary series of promotions. There are a couple of reasons. First, these positions are scarce and a suitable one is statistically more likely to occur somewhere else in the industry than at your current company. Second, your current colleagues know your weaknesses as well, if not better, than your strengths, and stepping up from the ranks to supervise your current peers is not the easiest undertaking. You will probably have to change employers to make the upward move that you seek.

Management is not for everybody, and accurate self-awareness pays when you assess your aptitude for it. Status and increased compensation can seem attractive, but how will you fare at providing leadership and example? Will you have the patience for constant meetings, arguments about budgets and occasional efforts to shift the blame for a setback onto yourself and your co-workers? If colleagues in your group disagree, will you be the diplomat who negotiates peace? Every employee owes their company proper loyalty, but working at the Director and Vice President levels requires you to implement directives from the company's senior levels however you feel about them personally. This can be seriously difficult when it includes informing hard-working people with whom you have shared laughter and success that they are about to lose their jobs, even though they are not at fault. Nor is everyone equipped with the toughness that it takes to discipline or, in the ultimate case, fire the employee who repeatedly fails to meet his or her obligations.

You will also receive pressure and hard questions about your group's performance. Can you give the company a positive, fact-based opinion of its contribution, and will you be able to inform the group of what it needs to do while convincing it that optimism is in order and the future is bright?

Then there is the question of workplace safety, especially for laboratory researchers. Will you insist effectively that company policies are followed, not being either lax or officious, but persuading all colleagues to create a high standard and hold their peers to it? This is a huge responsibility, as it is hard to imagine a worse situation than having a co-worker under your direction receive a serious injury. Apart from personal anguish, you will be involved with the legal ramifications that follow. Managers should remember that the literal meaning of "negligence," the usual charge against a company in these circumstances, is "failing to take care."

In any supervisory position, comply with all of your company's requirements on the safety front. Build a documented record of active surveillance of the safety of your workplace, such as regular inspections. If something goes amiss, even without causing an injury or lasting damage, ensure that the record shows that you inquired into the incident and took every reasonable measure to avoid a recurrence.

Try hard to build a culture in which any employee, and particularly any supervisor, speaks up immediately when somebody has violated workplace safety rules. Finally, if you join a small company that has yet to build a strong safety culture, park everything else until you have seen an improvement. You will be doing the right thing.

Companies and staff need good leaders who can manage well in both upward and downward directions. Taking on the responsibility has to be a good thing for both them and yourself – otherwise, you should leave it to somebody else who has the right gifts.

The management and leadership of groups are not quite the same thing, no matter what business school jargon says. It is not just cheerleading, but reflects reality, to state that everyone in a

Chapter 13: Should I Stay or Should I Go? Moving Into Management

scientific group should take their turn to lead according to the flow of the work. And where do leaders of scientists look for their inspiration? Robert Oppenheimer at Los Alamos – we set aside the terrible purpose of the work there – was fallible in his personal life, but his ability to be interested in every aspect of the site's project and to master the essentials of every corner of it, proved inspirational to the greatest collection of research talent ever united around a single task. (As an aside, the capacity of General Leslie R. Groves, as blunt and hard-charging an Army man as ever pulled on boots, to recognize that the ultra-academic Oppenheimer was the proper and only man for this job must rank as the greatest piece of human resource management in the history of that dark art).[58]

Some of Oppenheimer's decisions stand out. The brilliant suggestion of one scientist that nuclear detonation might follow explosive implosion of a hollow sphere of plutonium initially led Oppenheimer to put that man in charge of reducing the plan to reality. When his efforts proved unsatisfactory, Oppenheimer eventually replaced him with the individual whose skills best suited the problem, despite the resentment that this caused. The project became successful. No manager will enjoy making these tough decisions, but the capacity to deliver them when they are needed is a vital asset.

According to one account, Oppenheimer found this decision difficult and waffled about it for some time.[59] I see that as partly to his credit. A manager who makes important decisions without paying a cost of care is probably too hasty to be sound.

A second, more nuanced feat of Oppenheimer's was to keep Edward Teller onside as a useful member of the Project. Teller, an emotional, warmhearted and ambitious genius, was angered by his exclusion from a leadership role and burned to proceed with work on a topic beyond Oppenheimer's immediate concerns.[60] Oppenheimer gave Teller leeway – they were civilians, after all, and not Army officers – but received in return Teller's agreement to contribute in areas where he was particularly required. Variations of this form of management are sometimes needed in a company setting when a manager wants to retain a particularly talented employee with independent ideas.

In biopharma, only a small minority of ideas given the first steps of research attention lead to a marketed product. This creates a special need in both corporate and research management for a tolerance of frequent failure that must be married nonetheless to a steely determination to succeed.

Over recent decades, the industry has constantly examined its own processes and modified them in ways that seemed well-chosen at the time. One example relates to frequent expensive failures of seemingly promising medicines in Phase 2 of clinical trials, when the medicine is tried out in patients affected by the targeted disease. Hard questions about these failures led to one eclipsing all others – did the drug molecule even interact with its intended biological target? When the answer, too often, was "we don't know," it became a priority to make sure that it did.[61] Other related criteria were defined that became an essential part of any such trial.

Drug discovery is a hard game to win. Its purpose is noble and the long-term benefit to society worth working for, but its waters are choppy. Sail them with care.

Science-Based Employers Beyond Biopharma   14

The global economy ruthlessly creates downward pressure on prices. Adequate clothing is next to free thanks to the small wages paid to the overseas workers who produce it and the technology at their disposal. At the grocery store, few of us ignore the special offers of the week, even as we realize that the "savings" help to account for the low wages paid to workers at the deli counter. It takes steely virtue to choose fair trade coffee purchased directly from growers over cheaper mass-market brands. Only exciting innovations backed by persuasive marketing seduce us into paying more than we want to – I'm thinking of Apple's iPhones – and even then, competitors that some rate equal to or better than the original quickly appear at attractively lower prices.

Among the biggest actors in the stock market are pension funds whose investment portfolios finance the retirement benefits of teachers, state employees, and the like. Their managers' duty is to make money, and there is little scope for idealism beyond occasionally bowing to a strong imperative such as disinvestment from the tobacco industry.

Publicly traded stocks are valued according to the earning power of the issuing companies, and especially their anticipated future earnings. There is no mystery about this. If you tire of the lab one day and decide to start a neighborhood landscaping company, you will judge your success mainly by your profit-loss statement. The bank that lent you start-up funds expects you to earn enough to repay the loan with the agreed interest. If you do not, it will repossess your lawn mowers. Even to socially responsible investment managers, all the publicly traded biopharma companies represent the goods in just one aisle of the stock supermarket. If research-based organizations do not make money at a rate that satisfies Wall Street, investment managers will sell their stocks and invest in banks, airlines, technology companies, telecoms, real estate investment trusts and whatever else offers better financial performance. Sentiment isn't a factor, and little sympathy is spared for the weakened research-based biopharma that becomes subject to takeover and the cutbacks ("cost-savings") that follow.

While this challenge has been evolving, biopharma companies have had to climb an ever-steeper hill to get new medicines approved. Their greatest cost burden is already the salaries and benefits of staff, but internally sustaining all the functions required for successful drug discovery would require expanding this even further, and reduce profitability. Their response has been to restrain costs by outsourcing important functions to external companies that base their businesses on delivering the needed support.

Some of these organizations are overseas, but many are in the United States. They represent an important additional category of potential employers for the Young Scientist.

Major companies in this space include Charles River Laboratories (Wilmington, MA), Covance (Princeton, NJ) and IQVIA (Durham, NC) (IQVIA includes the former Quintiles). The services they provide range from experimental laboratory science over to clinical trial coordination and diagnostic services. Particularly for scientists interested in analytical methodology, they could be an interesting set of destinations. These are scientifically advanced organizations with a clear vision that cutting-edge developments offer them a huge future business opportunity.

In the past decade, there has been a trend toward the formation of strategic partnerships between major biopharma companies and larger contract research organizations (CROs). The attraction for the biopharma client is that large tranches of vital work can be entrusted to a stable partner, with sustained personal connections between managers on both sides improving the odds of mutual satisfaction.

For example, in 2010 Sanofi (then Sanofi-Aventis) entered into a 10-year $2-billion arrangement with Covance under which Covance became a "sole-source" supplier of central laboratory services to its partner. Signaling the depth of commitment made between the two organizations, Covance also took over two sites previously operated by Sanofi.

An even larger overall commitment from the biopharma side was separate arrangements made by Pfizer with a collection of CROs that included Parexel (Waltham, MA), Icon (Dublin, Ireland) and Pharmaceutical Product Development (known as PPD, Wilmington, NC). The apparent intention was to replace a greater number of project-specific agreements with outside vendors with durable partnerships covering multiple scientific and clinical functions.

Both large and small companies in this category can aspire to scientific leadership in selected fields, as they can assemble more expertise and operational capacity for a chosen type of task than a single biopharma can ever afford. Perhaps the best example is The Jackson Laboratory (Bar Harbor, ME), a not-for-profit institution that in some respects operates like a company and is the leader in providing transgenic animal models of human disease.

Skill- and technology-intensive activities like proteomics may best be handled by specialized groups in smaller companies that establish a protocol and then apply it every day to samples received from outside clients. This is fee-for-service work, but part of the service is to guide the clients in interpreting the results.

Inside a biopharma company, it can look like a luxury to invest the time necessary to optimize these exceedingly complex methods and then – a critical requirement – keep them running to ensure ongoing high quality. True expertise with a complex protocol is almost impossible to sustain if you do not practice the skills daily.

If your skill-set fits with the needs of a technology-based company, don't ignore the possibility of finding a career (or the beginning of one) there. It may give you a chance to continue working with methods that you acquired during graduate training that are considered interesting in biopharma but too resource-hungry or insufficiently mission-critical to be sustained in-house.

Call Me By My Brand Name (Until My Patent Expires)

15

Drug names are important for several reasons. The first is medical accuracy: when a physician prescribes a drug to a patient, it is clearly important that the patient receives the medicine that the doctor had in mind. Getting the wrong one is doubly damaging, because the intended treatment is never delivered and an unintended risk of harm is created by giving inappropriate medication.

Since 1992, the FDA's Center for Drug Evaluation and Research (CDER) has tracked the occurrence of medication errors in the United States and worked to reduce them. There are risk factors all along the delivery chain – a notorious one is misreading of poorly handwritten prescriptions – but the potential for confusion between similar drug names is a problem that CDER can attempt to minimize at its source, which is the naming process itself.

Before we take a look at some of the most important biologics and most innovative small-molecule drugs available today, let's consider how their names are put together to meet the needs of science, medicine and business.

The foundation of today's system is the **generic name** given to each drug, whether it is a small molecule or a biologic. According to the FDA's Use of Drug Name Terms Policy, the generic name is "an official or unofficial designation by which a drug is commonly available, unprotected by a trademark." Placing this name in the public domain allows anyone to refer to the chemical that corresponds to the drug without restriction, and this promotes clarity in biomedical discussion and prescription. Moreover, that name can continue to refer to the drug forever, no matter where or by whom it is manufactured.

Typically, though, as we saw in Chapter 3, a new drug first enters the market under the protection of a patent owned by the innovator company, and is marketed exclusively by the innovator group until patent protection expires. During that period, the costs of its discovery and development and of other research must be recouped, and a profit must be made to reward and encourage shareholders (Figure 2-1). The normal practice is for the company that launches the drug to give it a **brand name** distinct from the public-domain generic name. It is also, typically, simpler and "brighter" than the generic name so that the brand name assists efforts to promote awareness and availability of the drug ("ask your doctor if <insert brand name here> is right for you.") Brand names invariably carry an initial capital (Lipitor®, Epogen®, and the like), but the corresponding generic names are lowercase (atorvastatin, epoetin alfa) unless they start a sentence.

At first glance, drug generic names resemble the imaginings of a sci-fi screenwriter ("Captain, we have entered orbit around lovastatin 4"), but they are not so random. Generally, they are bolted together from short roots that denote the class of drugs to which a compound belongs while also making its name unique. For example, the generic names of the cholesterol-lowering statin drugs all end in *–statin*, those of angiotensin-converting enzyme inhibitors end in *–pril,* and those of antivirals usually end in *–vir*.

The World Health Organization approves International Nonproprietary Names for important candidate compounds, and widespread acceptance of these names by national regulators achieves a degree of global consistency. For some older drugs, different legacy names may continue to be used in different countries. For example, it probably causes tourists some difficulty that the active ingredient in Tylenol® (a brand name not used in

Chapter 15: Call Me By My Brand Name (Until My Patent Expires)

Europe) is called acetaminophen in the United States but paracetamol in many other countries.

The names of drugs in the statin class (Table 15-1) achieve both individuality for each compound and collective identity for the set. Statins inhibit the enzyme HMG-CoA reductase (HMGCR), which catalyzes the conversion of 3-hydroxy-3-methylglutaryl-coenzyme A to mevalonic acid, an important step in the synthesis of cholesterol. As a class, they descend from a natural product called mevastatin discovered in Japan by A. Endo during the 1970's. Mevastatin exhibited some toxicity along with potent inhibition of HMGCR, but excitement about the potential of its mechanism led to further work at Merck & Co. in the United States that uncovered the compound now called lovastatin. Lovastatin, under the brand name Mevacor®, became the first member of this drug class to be marketed following its approval by FDA in 1987. Subsequent cycles of clinical experience and scientific refinement led to today's range of statins.

Statins differ in potency. Atorvastatin and rosuvastatin, two of the more potent and widely used agents, emerged later in the evolution of the group. Physicians have engaged in trenchant arguments about the value proposition represented by the availability of multiple medicines in this and other classes, and the derisive term "me-too medicine" has been applied to drugs approved after the prototype of a particular class. This criticism ignores the exhaustive nature of the discovery/development process, under which it would take a Soviet-style monopoly system to arrange for only one research program to operate in a given area. Patients appear to have benefited immeasurably from the hard-driving competition that brought multiple statins, HIV-1 protease inhibitors and hundreds of other modern drugs to the market place. In fact, one of several causes for regret in the consolidation of the industry is that it reduces the vigor of competition.

Many of these medicines will be inexpensive and medically valuable a hundred years from now, which from a coolly economical point of view seems like a good deal between patent-holders and society. If medicines are expensive for 10-15 years after they appear, but then take a steep and permanent drop in price, should this not be agreeable to everybody? [62] But people are not inclined to be coolly philosophical when it is their child's health or their parent's or their own that is at risk while a medicine is unaffordable because a drug company needs to satisfy Wall Street and wishes to fund its own future health. As we noted in Chapter 2, different countries address this sensitive issue in their own ways. The research scientist, like any citizen, can have his or her own views: here we are only reckoning with the impact of market forces on the employment of scientists.

Dual naming of small molecules (Brand®/generic) caters well to the system under which generic copies of a branded medicine enter the marketplace when the period of exclusivity defined by patent protection has expired. As we noted in Chapter 3, the first generic form of a medicine to be approved receives six months of limited exclusivity during which no other generic (apart from one offered by the innovator company, if it chooses) can be sold. After that, the market opens to all approved generics and the presence of multiple competitors offering the same medicine with the same generic name generally leads to a dramatic fall in the price of the drug.

The brand name of the product as originally marketed remains the property of the innovator company, and the branded version usually continues to be sold. It might have a competitive advantage over generic copies if patients (consumers) picked their own version of the medicine, but most people benefit from a health insurance policy under which their insurer covers much of the expense. Insurers must control the costs of the programs that they deliver, and care little for the feel-good benefits of branding.

Chapter 15: Call Me By My Brand Name (Until My Patent Expires)

Table 15-1. The Statins – Important Cholesterol-Lowering Agents

Brand name (Company)	Generic name	Approved by FDA	Structure	Comment
Mevacor® (Merck)	lovastatin	08/31/1987		First statin approved: a natural product: lactone ring is hydrolyzed to yield active form
Pravachol® (Bristol-Myers Squibb)	pravastatin sodium	10/31/1991		Note the close resemblance to lovastatin after lactone hydrolysis
Baycol® (Bayer AG)	cerivastatin sodium	06/26/1997		Withdrawn in 2001 after being linked to 31 deaths
Zocor® (Merck)	simvastatin	07/10/1998		As with lovastatin, hydrolysis of lactone after ingestion yields the active drug
Lipitor® (Pfizer)	atorvastatin calcium	12/17/1996		Total sales have exceeded $150 billion
Crestor® (AstraZeneca)	rosuvastatin calcium	08/12/2003		Highly potent
Altoprev® (Aura Labs)	lovastatin (ext-release)	06/26/2002	as above	Formulated for once-a-day dosing
Livalo® (Kowa)	pitavastatin (calcium salt)	08/03/2009		Useful to some patients who experienced side effects with other statins
FloLipid® (Tcg Fluent Pharma)	simvastatin (suspension)	04/21/2016	as above	Liquid formulation of simvastatin
Nikita® (Lupin)	pitavastatin (sodium salt)	08/04/2017	as above	Alternative formulation of pitavastatin

Lipitor® (atorvastatin) has been a hugely successful prescription medicine, but a plot of its sales from 1998-2016 shows the drastic impact created by patent expiration and the advent of generic copies (Figure 15-1.) The milder downward trend prior to patent expiration can be attributed to the market entries of new branded competition and generic forms of statins older than Lipitor®.

So what about biologics? Are protein drugs, whether antibodies, soluble receptors, hormones or growth factors as capable as small molecule drugs of being recreated by companies that follow the disclosures of innovators? Can they be given a second life as cheaper generics? The answer is yes, but with complications that have slowed the arrival and uptake of *biosimilars*, as generic versions of biologics are known.

Typical small-molecule agents have a precisely defined chemical structure that chemists can reproduce, and other key properties of the original drug – the rate at which its tablets dissolve in water, for example – can be emulated to satisfy regulators that patients using the generic form will receive substantially the same dose and effect as those using the brand version. In contrast, a monoclonal antibody has a molecular weight of 150,000 and consists of a population of molecules in which variation is present. The most predictable site of variation is in the small sugar chains – the N-glycosylation – attached to each of the two heavy chains of an immunoglobulin G-class antibody. Other protein drugs, such as erythropoietin, are more heavily glycosylated and acquire even more diversity in the same way. For this reason, the standard by which biosimilars are assessed for their structural correspondence to original drugs has to be more elastic than the one for small molecules.

Despite this difference, the Brand®/generic naming system is used for biologics in the same way as for small molecules. As the number of monoclonals coming to the clinic increased, the United

Chapter 15: Call Me By My Brand Name (Until My Patent Expires)

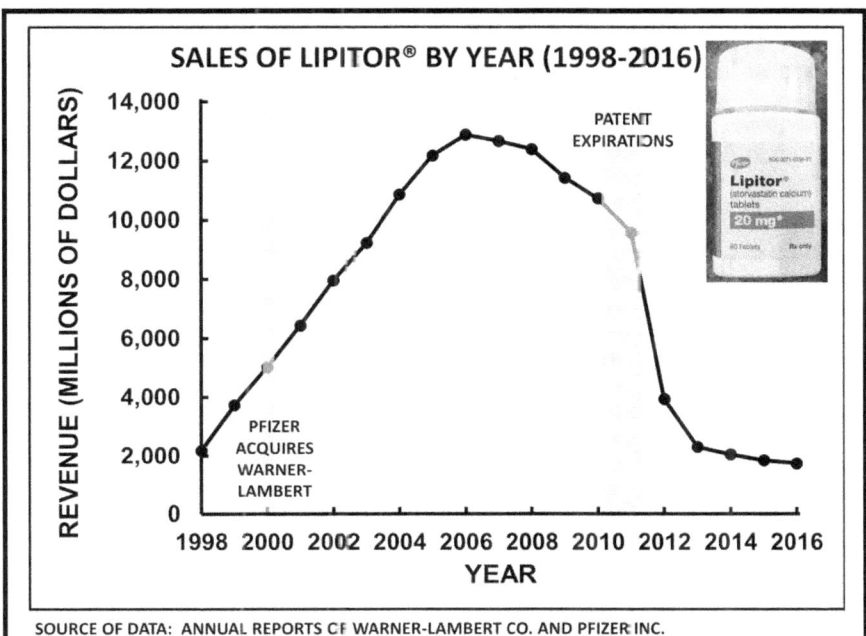

Figure 15-1. The dramatic effect on sales of a major brand name drug following loss of exclusive marketing rights.

States Adopted Names (USAN) Council adopted a systematic set of stems and suffixes to assist with the construction of suitable generic names for them. For drugs of all types developed in the United States, USAN coordinates the selection of suitable generic names and makes recommendations to WHO for their adoption as International Nonproprietary Names (INN).[63]

For antibodies, names are assembled from four components:

1. Prefix – unique to the individual molecule being named.
2. Infix (or substem) A – denotes the targeted biosystem or disease.
3. Infix (or substem) B – indicates source of the molecule.
4. Stem – the suffix –*mab* indicates a product containing an immunoglobulin variable domain that binds a defined target.

For example, Lucentis® (generic name: ranibizumab) is an engineered fragment of an anti-VEGF (vascular endothelial growth factor) monoclonal antibody used to treat the eye disease macular degeneration.

1. The first, unique part of the name, the single letter "r", distinguishes ranibizumab from other agents in the same category.
2. The first infix "anibi" signifies that the drug is an inhibitor of angiogenesis (blood vessel development).
3. The second infix "zu" indicates that the antibody sequence has been humanized.
4. The suffix "mab" indicates that it is a monoclonal antibody (or, in this case, a fragment in which the antigen-binding site is preserved).

This system also allows extra features (e.g. a linked toxin or polymer chain) to be indicated by adding on the name of the linked entity. For example, the antibody–drug conjugate Mylotarg® is known generically as gemtuzumab ozogamicin. It is used to treat CD33-positive acute myeloid leukemia.

Antibodies named before the introduction of systematic naming were not renamed, so inconsistencies can be found in the names of older molecules.

Marvelous Medicines.
Part 1: Biologics

16

"The lines between innovator and generic companies or between pharmaceutical and biotechnology companies have become increasingly blurred, and most major multinationals now incorporate both biologics and generics subsidiaries in their portfolios" – US International Trade Administration, 2016.[64]

The young biotechnology industry of the late 1970's quickly realized that clinical development of its products required partnerships with deep-pocketed pharma companies, but traditional pharma did not see at once that biologics would come to approach small-molecule medicines in importance.[65] It also believed for some time that patients would always prefer an orally active small-molecule drug to an injectable. Another reason for hesitation was that monoclonal antibodies, despite their obvious potential as "magic bullets," proved harder than expected to turn into drugs.[66]

The emergence of massively successful biologics changed minds eventually,[67] and some prominent biotechs (e.g. MedImmune, Genentech, Genzyme) were absorbed by traditional pharma companies that came to foresee a future based equally on large- and small-molecule drugs. Today, Deng Xiaoping's famous aphorism about economics also applies to drugs – "it doesn't matter if a cat is black or white as long as it catches mice." Physicians have learned not to care much whether a medicine is based on a large or small molecule if it delivers benefit to the patient.

Biologics now command around 25% of the market for medicines,[68] and include seven of the world's ten top-selling medicines (Table 16-1). It's instructive to look at world-leading biologics to consider their mechanisms of action and to see why they are so medically important.

As an aside, vaccines represent a separate category, but their potential to liberate massive populations from feared diseases give them huge medical and economic value. Use of recombinant DNA technology to generate multivalent vaccines that immunize against multiple epitopes of pathogens has boosted their potential, broadening the range of conditions that can be prevented. Classical vaccines include those against polio, smallpox, and MMR (measles/mumps/rubella combined vaccine). Important new vaccines immunize patients against herpes virus, shingles and pneumonia, while keen efforts to develop vaccines against malaria and HIV-1 continue with valuable support from the Gates Foundation.

Major biopharma companies, including GlaxoSmithKline, Merck, Sanofi and Pfizer are active players in vaccines, with smaller companies such as Novavax (Gaithersburg, MD), Emergent BioSolutions (same location) and VBI Vaccines (Cambridge, MA) equally attracted by the exciting potential that vaccines offer for business and medical advance.

Mainstream protein biologic medicines exhibit three principal modes of action: **replacements, interceptors** and **biological signaling agents**.

Replacements: Correction of a chemical deficiency is an established concept in medicine. Orally dosed small-molecule supplements are recognized treatments for important pathologies such as Addison's disease (cortisone analogs), hypothyroidism (synthetic thyroid hormone), Parkinson's disease (L-DOPA), post-menopausal symptoms in women (estrogens), or frailty in men (testosterone). Some protein biologicals work on the same principle, but must be administered by injection.

Chapter 16: Marvelous Medicines. Part 1: Biologics

Insulin derived from animals and growth hormone derived from human cadaver pituitary glands were used for many decades to treat diabetes and short stature, respectively. The development of recombinant DNA-based protein expression immediately offered the prospect of replacing these tissue-derived preparations with genetically engineered polypeptides. Medical expectations included avoiding immune responses to animal insulin and elimination of the occasional occurrences of prion-mediated Creutzfeldt-Jakob disease caused by using human brain tissue extracts. Today, improved and refined versions of these prototypical agents command markets in excess of $1 billion that signify their life-saving and life-enhancing medical value to large numbers of patients. A slow-release form of insulin (Lantus®, Novo Nordisk) was until recently among the world's ten top-selling medicines, with 2018 sales in excess of $4 billion.

Erythropoietin (EPO) is a 20 kDa protein hormone produced in the kidneys that stimulates the production of red blood cells in the bone marrow. EPO production suffers in cases of chronic renal disease, while the bone marrow is injured by chemotherapy for cancer. In either case, supplementation with recombinant EPO helps to restore red blood cell production. EPO has also been a notorious drug of abuse by some athletes in endurance sports, especially professional cycling.[69]

Other important replacement therapies are the recombinant-derived clotting factors used by hemophiliacs, and granulocyte colony-stimulating factor for low white-blood cell count (cytopenia).

Manufacturers provide websites based on product brand names that provide a blend of promotional and medically objective information for the benefit of patients and providers.

The medicines mentioned above are used to treat chronic diseases, but recombinant-derived tissue-type plasminogen activator (t-PA) is used in thrombotic disease emergencies. It came to market in

1987 as an acute intervention in heart attack, but has also found value as an intervention in stroke. It is often called a "clot-buster," but t-PA is actually an activator of the clot-busting fibrinolytic system. Its proteolytic action on plasminogen activates that more abundant proenzyme to plasmin, and it is plasmin that "busts" the clot by cleaving fibrin to reopen an occluded blood vessel in the heart or brain. The therapy comes with the risk of creating excessive intracranial bleeding, and needs to be administered within a few hours of the onset of the crisis.

Protein drugs administered by injection reach their sites of action through the bloodstream. The low molecular masses of some – insulin is a 6 kDa polypeptide, EPO 20 kDa and somatotropin (human growth hormone) 22 kDa – fall below the permeability threshold of the kidney glomerular membrane and cause them to have short circulating half-lives. The glomerular membrane retains larger plasma proteins such as albumin (66 kDa), serotransferrin (80 kDa) and IgG (155 kDa) in the plasma, but smaller proteins can pass on to the renal proximal tubules, where they are reabsorbed and degraded.[70] This can be countered by chemically modifying smaller biologics with polyethyleneglycol (PEG), a relatively inert polymer (Figure 16-1). The extended lifetimes of PEGylated forms of some early biologics have allowed them to displace the originals as market leaders.

Interceptors: Immune self-tolerance, a topic studied by Sir Peter Medawar, is the body's programming against mounting an immune response to its own cells and tissues. Failures of self-tolerance result in autoimmune diseases. A prevalent example is rheumatoid arthritis (RA), which begins with painful and swollen joints and can proceed to damage other organs, including the heart. Inflammation is a complex response to injury and infection that causes redness (hence the name), swelling and pain, and can be moderated by a number of drugs. Among these, the small-molecule agent methotrexate is widely prescribed.

Chapter 16: Marvelous Medicines. Part 1: Biologics

Table 16-1. Ten Top-Selling Pharmaceuticals in 2017 (Global Sales)

	Rank	Drug	Molecular type	Treats	Vendor	Action	Sales ($10⁹)
✓	1	Humira®	monoclonal Ab	rheumatoid arthritis, Crohn's disease, ankylosing spondylitis, psoriatic arthritis, plaque psoriasis, ulcerative colitis	AbbVie	interceptor of TNF-α	18.4
✓	2	Rituxan®	monoclonal Ab	non-Hodgkin's lymphoma, chronic lymphocytic leukemia	Roche (Genentech)	anti-CD20	9.2
	3	Revlimid®	small molecule	multiple myeloma	Celgene	induces degradation of IKZF1 and IKZF3	8.2
✓	4	Enbrel®	soluble receptor	rheumatoid arthritis, plaque psoriasis, psoriatic arthritis, ankylosing spondylitis, juvenile idiopathic arthritis	Amgen	interceptor of TNF-α	7.9
✓	5	Herceptin®	monoclonal Ab	HER2⁺ breast cancer w/ spread to lymph nodes; HER2⁻ breast cancer with certain conditions	Roche (Genentech)	binds to HER2 receptors	7.4
	6	Eliquis®	small molecule	reduction of risk of stroke in nonvalvular atrial fibrillation; reduction of risk of dep vein thrombosis	BMS/Pfizer	factor Xa inhibitor	7.4
✓	7	Remicade®	monoclonal Ab	rheumatoid arthritis, Crohn's disease, ankylosing spondylitis, psoriatic arthritis, plaque psoriasis, ulcerative colitis	Janssen Biotech (J&J)	interceptor of TNF-α	7.2
✓	8	Avastin®	monoclonal Ab	colorectal cancer, other cancers	Roche (Genentech)	interceptor of VEGF-A	7.1
	9	Xarelto®	small molecule	reduction of risk of stroke in nonvalvular atrial fibrillation; reduction of risk of dep vein thrombosis	Bayer/J&J	factor Xa inhibitor	6.6
	10	Eylea®	soluble receptor	macular degeneration (wet); macular edema	Bayer/Regeneron	interceptor of VEGF-A and VEGF-B	6.0

Source: Genetic Engineering and Biotechnology News (GEN)   ✓ often co-dosed with small molecule agent

117

Cytokines convey a refined set of messages between cells of the immune system. They exist at low concentrations, but act in the same way as classical protein hormones. A signal originating at one type of cell is delivered by the cytokine to another through a mechanism that includes binding to a receptor at the target cell's surface.

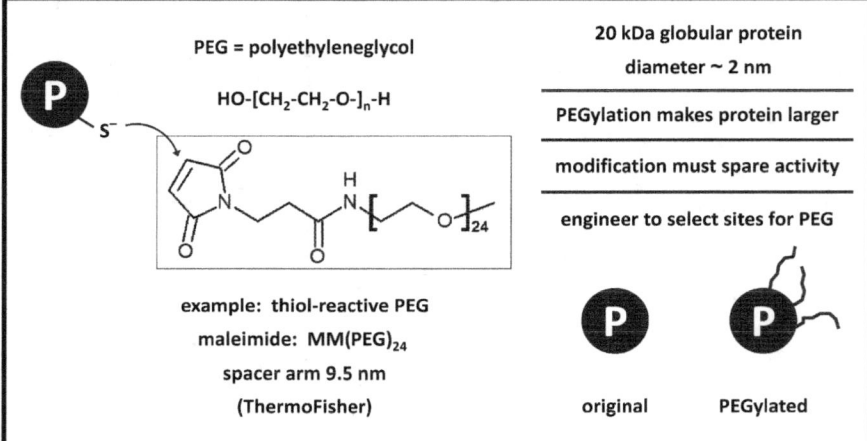

Figure 16-1. Modification of a protein thiol group by a maleimide derivative of polyethyleneglycol (PEG). Right: Enlargement of protein radius by pegylation. PEG chains must be attached site-specifically to avoid interfering with the protein's biological activity.

Proinflammatory cytokine drivers of RA include tumor necrosis factor-α (TNF-α) and interleukin-6 (IL-6).[71] Interceptors that bind TNF-α to suppress its signal have become enormously important medicines, and their high medical value in a large patient population has caused them to top the tables of revenue earners during their periods of market exclusivity (Table 16-1), and to attract the attention of biosimilar producers seeking a share of these markets.

Two different types of protein drug both serve as TNF-α interceptors, monoclonal antibodies (mAbs) and soluble receptors.

In 2017, the top-selling drug in the world with $18.4 billion in revenue was Humira® (generic name: adalimumab), an anti-TNF-α mAb marketed by AbbVie (North Chicago, IL). In seventh place was Remicade® (generic name: infliximab) from the Janssen Biotech unit of Johnson & Johnson (New Brunswick, NJ), also an anti-TNF-α mAb, with sales of $7.2 billion. The potential of mAbs to be used as interceptors of disease-driving proteins was imagined to be vast from the time of their invention by César Milstein and Georges J.F. Köhler in the mid-1970's, but, as we noted above, unforeseen challenges took time to overcome.

The second type of agent used to trap or "sop up" TNF-α is represented by Enbrel® (generic name: etanercept) from Amgen, which accounted for global sales of $7.9 billion in 2017. Etanercept is a soluble receptor, a molecule created by engineering the extracellular ligand-binding domain of the TNF-α receptor into a dimeric structure that binds the cytokine (Figure 16-2). As with

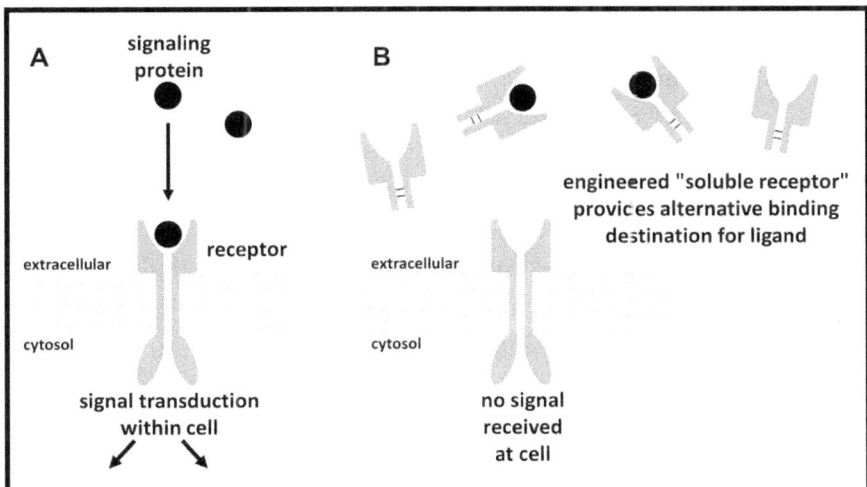

Figure 16-2. Soluble receptor functioning as an intercepting agent. (A) Binding of a signaling protein to its membrane-bound receptor delivers a signal to the receiving cell. Excessive levels may be implicated in a disease state. (B) An engineered soluble form of the receptor is administered as an interceptor and prevents the signal from being delivered.

the mAbs, administration of this construct introduces a soluble competitor for TNF-α that reduces the amount reaching the natural cell-bound form of the receptor and thereby interferes with the signaling function of the cytokine.

Most drugs are developed entirely in biopharma companies, but etanercept was devised by researchers at the University of Texas Southwestern Medical Center in Dallas.[72] It was licensed to Immunex, a Seattle-based biotech with world-class research capacity in cytokine biology. Immunex was acquired by Amgen in 2002 for $16 billion.

Treatment with small- and large-molecule agents is not an either-or proposition. Patients frequently receive a combination of both as their physicians see fit (Table 16-1). The complementary actions of multiple drugs add up to form an optimized overall treatment.

The use of biologic drugs as interceptors of signaling proteins has become powerful and prevalent in medicine. Identification of the connections between individual cytokines and particular forms of inflammatory disease has allowed more of them to be targeted, mainly with mAbs (Table 16-2).

Recognized inflammatory disorders include RA, osteoarthritis, various forms of psoriasis, ulcerative colitis and Crohn's disease. Beyond this incomplete list, it is known or suspected that dysregulated inflammatory processes contribute to a broader spectrum of pathologies including neuroinflammation in Alzheimer's disease, vascular injury in cardiovascular disease, and tissue damage in autoimmune diseases such as multiple sclerosis and systemic lupus erythematosus. Hypothetically, there should be scope to treat these conditions with cytokine interceptors if the proper target can be selected. This potential reflects perfectly the way in which great basic research in academic institutions opens the road for life-changing advances in drug discovery.

Table 16-2. Additional Monoclonal Antibody (mAb) Interceptors of Cytokines

Brand name	Drug	Target	Treats	Vendor
Taltz®	ixekizumab	interleukin-17A	psoriasis	Lilly
Cosentyx®	secukinumab	interleukin-17A	psoriasis, ankylosing spondylitis	Novartis
Siliq®	brodalumab	interleukin-17A receptor	psoriasis	Valeant
Tremfya®	guselkumab	interleukin-23 (p19 subunit)	psoriasis	Janssen Biotech (J&J)
Stelara®	ustekinumab	interleukin-12 / interleukin-23	Crohn's disease	Janssen Immunology (J&J)
Fasenra®	benralizumab	interleukin-5 receptor α-chain	severe eosinophilic asthma	AstraZeneca
Ilaris®	canakinumab	interleukin-1β	cardiovascular disease (emerging)	Novartis

A promising result of this type appeared in 2017, when the anti-interleukin-1β mAb canakinumab was shown to reduce significantly the number of recurrent cardiovascular events in patients who had previously suffered a heart attack.[73] Novartis, the trial sponsor, announced in October 2018 that FDA had not granted approval for the new indication, but we can expect more exciting forays into new medical territory using biologics.

Canakinumab's effects on heart disease were not delivered through effects on patients' plasma lipid levels, but biologics have also entered that area in recent years. In patients who do not respond adequately to treatment with diet, exercise and statins, mAbs directed against the circulating protein PCSK9 are used to lower circulating levels of LDL-cholesterol (cholesterol in low-density lipoprotein particles, also called "bad" cholesterol). PCSK9 binds to the LDL receptor (LDLR), the cell-surface protein that pulls LDL particles from the blood into cells, and causes the receptor to be degraded rather than recycled. Therefore, PCSK9 tends to elevate plasma levels of LDL-cholesterol. Intercepting PCSK9 with a mAb can deliver dramatic cholesterol-lowering effects. Currently marketed anti-PCSK9 mAbs are Repatha® (generic name: evolucumab) from Amgen (Thousand Oaks, CA) and Praluent® (generic name: alirocumab) from Sanofi (Gentilly, France). Praluent® was discovered by Regeneron (Tarrytown, NY) and developed in partnership with Sanofi.

PCSK9 interceptors are the fruit of modern methods of genetic analysis. Starting in around 1970, Michael Brown, Joseph Goldstein and their many research colleagues successfully probed the question of why about 1 in 500 individuals is affected by the heterozygous form of familiar hypercholesterolemia (FH), an inheritable condition in which the plasma LDL cholesterol level is double the normal value.[74]

Molecular cloning methods, which at the time were the very

cutting edge of biological research, led them to identify two genes in which mutations could cause FH. These were the genes encoding the LDLR and the low-density lipoprotein component apolipoprotein B (ApoB), respectively.

When physicians encountered FH patients who lacked any abnormality in LDLR or ApoB, a search began for additional genes that are crucial to proper cholesterol management. Population genetic analyses by a group led by Helen Hobbs, a former postdoctoral fellow of Brown and Goldstein, showed that certain variants of the gene *PCSK9* were correlated either with reduced risk of cardiovascular disease or with increased risk.[75] With the discovery that even heterozygous loss of function of the *PCSK9* gene product aligned with substantial reductions in the risk of disease, a clear case could be made that a drug that blocked the function of this protein would produce a similar effect. So it proved, and Repatha® and Praluent® are the fruits of this insight.

The PCSK9 story represents an almost ideal case of "target validation" – an indication that "drugging" a certain target will lead to a physiological response – because both positive and negative effects on a health-related biomarker can be detected. In its dreams, the biopharma industry would receive this kind of pathfinding information for every prospective target, but it will only happen in relatively rare cases.

Dr. Judah Folkman (1933-2008), a famed pediatric surgeon at Boston Children's Hospital, is credited with advancing the theory that angiogenesis – the development of new blood vessels – is a crucial step in the development of solid tumors.[76] His ideas launched biochemical searches for protein signaling factors that stimulate this process. Today, mAbs or soluble receptors that intercept VEGF are used as both cancer treatments and also in the important eye disease age-related macular degeneration (AMD). Avastin® (generic name: bevacizumab), discovered by Genentech and marketed by its parent, Roche Holding AG (Basel, Switzerland), is an important treatment for solid tumors such as those of the brain, kidney, colon, rectum or lung on the premise that VEGF is an important actor in supporting their growth. An engineered fragment of the same antibody is marketed as Lucentis® (generic name: ranibizumab, also mentioned in Chapter 15) to block abnormal new blood vessel growth in the macula, and competes in that market with Eylea® (generic name: aflibercept) from Regeneron, an engineered soluble receptor.

The antitumor agent Avastin® targets the same growth factor as the eye drug Lucentis® by the same mechanism, but is many times cheaper in terms of dollars per microgram. As a result, ophthalmologists often reduce the cost of treating macular degeneration by using Avastin® "off-label" to treat their patients with AMD. This means prescribing an approved medicine for a therapeutic purpose other than the one for which it was approved. Physicians have broad freedom to do this, but biopharma companies are not permitted to promote off-label uses. Compounding pharmacies perform the subdivision of Avastin® and supply it to the physicians. This rather controversial practice departs from the established paradigm that medicines move directly from the strictly quality-controlled production line of a biopharma manufacturer into the hands of the medical practitioner who administers it. The argument in its favor is the cost-saving that it affords.

The concept of molecular interceptors as drugs is a beautifully simple strategy that capitalizes on insights from basic biology. As we have seen, it is already delivering benefit to patients affected by inflammatory diseases, cancer, eye disease and hyperlipidemia, and further successes can be expected.

Signaling Molecules: Our third important class of biologics is molecules that deliver a biological signal in order to modify an ongoing process associated with disease. Together with insulin analogs, a second important class of peptide agents used to treat diabetes is the incretin analogs. Incretins are produced by the gut when food is taken and stimulate the secretion of insulin by the pancreas. They also slow down the rate at which the stomach empties itself into the gut, and thereby moderate the rate at which the system submits food to the main phase of digestion. Byetta® (exenetide), its extended-release version Bydureon® (both from AstraZeneca, Cambridge, UK), and Victoza® (liraglutide: Novo Nordisk) are derivatives of the natural glucagon-like peptide-1 (GLP-1).

Diabetics and their doctors can choose from a wide range of therapeutic options, including small-molecule and peptide-based agents. GLP-1-related peptides are not a form of insulin, and are not suitable for type 1 diabetics, whose bodies do not make insulin. Multibillion dollar sales indicate their value in type 2 diabetes, a condition in which the body's tissues become less responsive to insulin.

The human immune system consists of two parts, innate immunity and acquired immunity.[77]

Innate immunity equips us to respond to threats that have faced multicellular organisms since they emerged, such as bacteria, viruses and tissue damage. These are detected by receptors that recognize characteristic biomolecules. For example, detection of the bacterial cell wall component lipopolysaccharide raises a biological alarm that summons a response from macrophages and other cells that constitute this system.

Acquired immunity is the capacity to respond to a completely new threat, and rests in the fact that the body maintains a standing army of B cells and T cells. Each of these cells displays one of a vast repertoire of antigen-binding receptors generated by cutting and splicing DNA from a number of genes. Foreign (but not host) antigens stimulate immune cells that carry an appropriate receptor to undergo clonal amplification, with B cells leading the humoral (antibody-based) immune response to foreign invaders, and T cells attacking host cells infected by pathogens.

Immune surveillance is a continuous process of inspection conducted by the immune system. Evolution has elected for cells of the body to "show their papers" by presenting fragments of the proteins that they make on their surfaces, where the fragments are carried by histocompatibility antigens. Tumor cells that express antigens recognized as foreign are, preferably, attacked and destroyed, but this battle is not a foregone conclusion. The immune system has to function with enough moderation to avoid indiscriminate damage to the body, and is reined in by having an element of negative signaling or "braking" built into its response. A recently evolved strategy to combat cancer is to suppress this negative signal so as to maximize the intensity of the attack mounted against the tumor.

The checkpoint protein PD-1 on activated T cells responds to binding a partner ligand called PD-1L (PD-1 ligand) that is found on many healthy cells. Binding of PD-1L to PD-1 sends a signal to the immune cell to discontinue or downregulate the attack that it would otherwise mount. PD-1L is also abundant on some tumor cells, which allows them to falsely pacify the immune system. Therapeutic mAbs that prevent this interaction from taking place are promising agents for the treatment of a range of cancers. An important example is Keytruda® (generic name: pembrolizumab), marketed by Merck and Co.[78]

A similar class of agents is anti-CTLA4 mAbs. CTLA4 is another protein expressed at the surface of T cells that, on binding a ligand, sends a message that downregulates the immune response. This moderating message can be abrogated by a mAb. Clinical experience with the two main examples of this type of therapy has been variable, with occasional complete remissions of cancer interspersed with less satisfactory results.[79]

In terms of the division of biologics into replacements, interceptors and signaling agents, the anti-PD1 and anti-CTLA4 mAbs could reasonably be placed in either of the last two categories. Unlike incretins, which deliver a positive signal, the antitumor mAbs act by interrupting a signal that might otherwise have been sent.

Immunotherapy of cancer has a long history. The New York physician Dr. William B. Coley (1862-1936) observed that patients bearing a solid tumor occasionally experienced dramatic remission after suffering from an unrelated infection. He experimented with immunization using the BCG vaccine, achieving occasional successes that may have been overshadowed by the emergence at the same time of radiotherapy.[80]

The development, marketing and prescription of biologics have developed rapidly over the past 30 years. There are many reasons for this, one of which is that advances in certain key technologies such as industrial-scale cell production and the humanization of mAb sequences are generic to the field and can be applied broadly. The undoubted medical value of biologics has made the high prices often charged for them acceptable so far, especially when other costs are being displaced. The sustainability of this trend remains to be judged.

Biosimilars (generic forms) of leading biologics have appeared, but in the U.S. have not yet achieved the deep price cuts typical of generic forms of important small-molecule agents. Reports indicate that in some other countries, where the government is essentially the sole purchaser of medicines, bidding wars between originator and emulator companies have driven some decisive shifts to biosimilars.

As it happens, there have not yet been many cases of large- and small-molecule medicines competing for the same niche. An exception is the area of rheumatoid arthritis and related inflammatory disorders, in which protein drugs used to block the master cytokine tumor necrosis factor-α face an alternative in the form of small-molecule drugs that inhibit protein kinases involved in inflammatory signaling. Curiously, it was the large-molecule agents that took early ownership of this important market, and the small-molecule agents that emerged later. We will examine this case of competition in Chapter 17 after reviewing some other recent advances in small-molecule therapeutics.

Marvelous Medicines.
Part 2: Small Molecules 17

The rise of biologics takes nothing away from the attractiveness of orally active small-molecule drugs. The ease with which they are prescribed, distributed and dosed remains an enormous positive. In recent years, however, the rate at which new pharmaceuticals are being discovered has been slower than medical and business leaders once expected and hoped.

Nevertheless, marvelous new medicines have been discovered and brought to market since the year 2000. Before considering some of them, let's digress briefly into discussing the system that supplies them.

As the Human Genome Project approached completion in the early 2000's, hopes were high that it would provide a bounty of targets for new drugs. In this particular respect, the result was disappointing.[81] Human protein-coding genes were surprisingly few in number (a recent estimate is 21,000,)[82] and the fraction encoding "druggable targets" was gauged to be only 10-14%.[83] No runaway train of drug discovery ever left the station. The pharmaceutical industry's productivity in the first decade of the new century lacked the vigor of its boom years in the 1990's.[84]

Reflecting on this downturn, the economic strategist M. Gittelman suggested that "the rise of basic science in medical research may be a setback for drug discovery." She suggested that the older paradigm under which medical innovation occurred in hospitals through feedback-based learning might have given better results.[85] This cannot be disproved, but the scientist whose career goal is active research in a company can consider the discussion above their pay grade for the first few years at least. Moreover, the high-level policies that shaped medical research for the past fifty years have by now moved too many mountains for their effects to be undone in a hurry.

Here are three of Gittelman's main points. First, the precedent set by twentieth century physics has not recurred in biology. Advances in physical sciences transformed technology, but the high complexity of human biology has frustrated an equivalent conversion in medicine. (Yes, there has been massive progress, but drug discovery continues to be a less tractable technology than landing on the moon or developing cellphones). Second, changes in medical practice, especially specialization and the savings-driven shortening of hospital stays, have lessened the physician's opportunity to observe patients at length and receive their feedback. Third, it is a reductionist error to anticipate that disease-causing lesions at the molecular and cellular level will correspond to the interventions needed to supply cures. In fact (says Gittelman), the detection by attentive physicians of unexpected drug actions has been as good a basis for useful new discoveries as science-driven predictions of clinical action based on molecular effects.

Against this background of supposed difficulty, biologics have emerged and flourished. As we saw, they mostly have relatively simple actions as replacements for or interceptors of natural protein factors. Small-molecule drugs, in contrast, have to penetrate the system and achieve controlled modulation of a target pathway while sparing all others. Achieving this remains an imposing challenge, but the accumulated experience of many decades can be drawn upon to meet it.

How many new drugs should be discovered every year? The question is meaningless in isolation. Context – and with it the notion of a shortfall – comes when we recall again how business and science intertwine in biopharma. The shortfall is less about science than about failure to achieve the scale of business success demanded by investors.

Consider things from the perspective of the pension fund manager seeking profitable investments. Preferably, the manager buys and holds shares in companies that grow in value year-over-year. Some companies, generally the larger ones with predictable revenues, may also spin off cash to shareholders in the form of a regular dividend.

When a smaller biopharma company has annual sales of $100 million, modest new sales of $10 million per year constitute a growth rate of 10% that will tend to be reflected by increases in its stock price. But when a major company has revenues of $20 billion per year, a growth rate of 10% requires bringing products that earn $2 billion of new revenue to market every year, which is a nearly impossible 1-2 major drugs per year. If productivity is at a lower level, the stock price of the major company is in danger of stagnating.

The pension manager can risk investing in the fast-growing small company, but will know that it is only one medical setback away from an abrupt reversal of fortune. He or she may also settle for the safer but less exciting performance of the larger company. But fast-growing companies in other areas of business represent competition for the investment dollars of the pension fund. A company whose stock drops to a low value is at risk of being taken over by a competitor or an opportunistic investment fund. This is why purely financial pressure is a main factor in determining the business choices of biopharma companies. A company cannot benefit society and its stakeholders if it does not first survive and prosper.

In addition to the challenges of new drug discovery, losses of revenue caused by existing drugs losing their patent protection mean that major biopharma companies grapple daily with a colossal challenge. Efforts to meet it (or come close) have driven extensive consolidation in biopharma through mergers and acquisitions over the past two decades. Commentators offended by the profits of surviving biopharma companies may not notice that less successful companies are disappearing rapidly from the landscape, often being cannibalized by the survivors.

Business pressures aside, there has also had to be a scientific response to changing views of how drug discovery should best be done. In the period that led to completion of the Human Genome Project, there was an industry-wide conviction that target-based methods of drug discovery were the way to go. Briefly, these consist of hypothesizing that drug-mediated intervention against a certain target protein will be medically safe and effective against a certain disease. Gene deletion experiments that emulate drug action might provide support, showing that a mouse can live without the function of a certain potentially targeted enzyme or receptor. Libraries of compounds are then screened for leads that display activity against that target in purified form and are later refined toward genuinely drug-like properties by immensely skilled medicinal chemists. This method has had its successes, but not at the level that would have matched the business needs of the industry's larger players.

Concern for the productivity of target-based approaches has revived enthusiasm for phenotypic screening, a *systems-based* rather than target-based approach in which an intact biological system, usually cells, is the platform for a screen that gauges the performance of a high-level function such as cell motility or the intracellular release of calcium ions. A phenotypic screen seeks compounds with activity against the biological function that we wish to inhibit or activate without foreknowledge of a molecular target. It is desirable, but not always essential, to determine the molecular target of the activity later.

Another apparent advantage of phenotypic screening is that the cell line employed can be chosen or engineered to emulate as closely as possible the disease condition that we ultimately hope to relieve.

This is not quite a return to physician-scientists collaborating with pharmaceutical industry scientists to make discoveries in the hospital, but it has an element of it.

In a widely cited 2011 paper, Swinney and Anthony[86] claimed that "the contribution of phenotypic screening to the discovery of first-in-class small-molecule drugs exceeded that of target-based approaches" by a count of 28 to 17 between 1999 and 2008. A look at this issue for the period 1999-2013 by authors from Novartis[87] reached somewhat contrary conclusions, but still recognized the meaningful fraction of drug discovery accomplished through phenotypic screens.

Target-based screening remains the dominant paradigm, and the evolution of science and technology opens new avenues for discovery to assist it. For example, nearly half of all known drugs target G protein-coupled receptors (GPCRs), but these transmembrane proteins were almost entirely beyond the reach of X-ray crystallography until 2007. New techniques that make them capable of being crystallized have led to a stream of informative structural results that seems sure to boost the numbers of drugs discovered against GPCRs by rational methods in the next 10-15 years and beyond.

Antibiotic discovery is an historical reference point for phenotypic screening. Over one hundred years ago, Paul Ehrlich and his colleagues at Bayer – valiant industrial researchers – sought new antibiotics by screening for bactericidal activity.[88] This is an area in which target-based discovery, despite being flooded with opportunities by new biology, has tended to be disappointing. One reason is that an antibiotic usually needs to kill organisms of several different species among which a particular enzyme target can show variation. Another is that the old problem of emergent resistance to antibiotics remains hard to overcome.

When new technology is allied to creative imagination, successful phenotypic screening can produce amazing results. A prime example is a new generation of small molecule drugs invented to benefit patients affected by cystic fibrosis, an inherited disease caused by mutations in a chloride channel known as the cystic fibrosis transmembrane conductance regulator (CFTR). Poor function of the mutant channel causes the lungs of patients to be clogged with mucus. This leads to impaired breathing, increased risk of lung infections and shortened lifespan. A large number of different mutations are known, and have a range of medical consequences.[89]

As a 1480-residue transmembrane protein, CFTR is unsuited to classical biochemical methods developed with soluble proteins. In addition, a drug that improves the function of defective forms of CFTR must display the end-action of improving chloride transport through the channel. It would be difficult to construct an assay for this activity using purified protein.

Assisted by "venture philanthropy" from the nonprofit Cystic Fibrosis Foundation (Bethesda, MD),[90] San Diego-based researchers of Vertex Pharmaceuticals (Boston, MA) applied a phenotypic screening system that originated in an academic collaboration.[91] It is based on cells that express mutant forms of CFTR and a halide-sensitive form of yellow fluorescent protein (YFP-H148Q). The fluorescence of YFP-H148Q is quenched by iodide, which properly functioning CFTR allows to pass as a surrogate for chloride. Using changes in the fluorescence of YFP-

H148Q as the readout, these cells were used to screen for compounds that enhanced iodide transport through otherwise poorly functional mutant channels, and compounds acting by two different mechanisms were discovered.

Compounds that interact directly with mutant CFTR forms that reach the cell surface but have poor channel function are called potentiators. Compounds that improve the ability of mutant CFTR to fold and reach the cell surface are called correctors. Vertex now markets Kalydeco® (ivacaftor) as a potentiator that significantly improves lung function in patients with any of several different mutations, but not (if given alone) for those with the most common mutation, ΔF508. These patients are helped by additional products, Orkambi® and Symdeko®, in each of which the potentiator ivacaftor is combined with a corrector.[92] Trikafta®, a three-compound combination that does benefit ΔF508 patients, was approved in October 2019.

Two advantages of phenotypic screening are that it maintains an intact cellular system and allows the discovery of compounds that target the system of interest by multiple modes. Hypothesis-driven discovery approaches are always limited to the current state of scientific knowledge, but phenotypic screening allows the discovery of compounds with unprecedented mechanisms. It also allows the detection of efficacy through polypharmacology, which is drug action delivered through multiple different targets. Polypharmacology has unexpectedly emerged as an important positive attribute of tyrosine kinase-inhibitor drugs that are important in oncology.

Small-molecule drugs are often coadministered with biologics (Table 16-1) so that both types of agent work together. There are not many cases in which they shape up as alternatives to each other, but one has emerged recently in the area of autoimmune disease. We saw that an important and growing family of biologics are the mAbs or soluble receptors that intercept proinflammatory cytokines. Unless intercepted, a cytokine delivers its signal by binding to its receptor on the surface of target cells, with the subsequent process of reading and acting upon the signal being termed signal transduction.

For cytokines, part of signal transduction is delivered by a family of protein kinases known as the Janus kinases, or JAK family. There are four of these enzymes, JAK1, JAK2, JAK3 and TYK2, and their activity of phosphorylating both the cytokine receptor and additional target proteins from the seven-member STAT family (signal transducers and activators of transcription) leads to a phosphorylated STAT protein being moved to the nucleus where it modulates the expression of target genes by binding to specific sites on DNA (Figure 17-1). Depending on the cytokine, the cast of receptor, kinases and STAT proteins varies, but the general mechanism is the same.

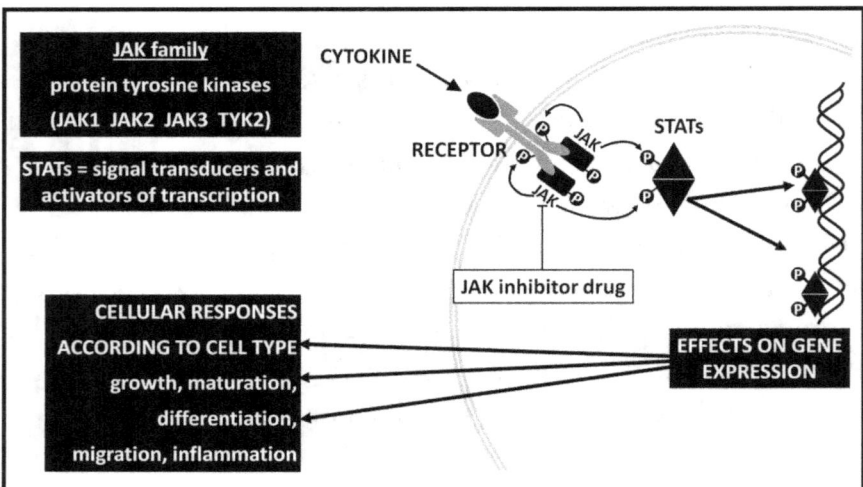

Figure 17-1. Signal transduction from a cytokine receptor to altered gene expression mediated by kinases of the JAK family and members of the STAT family of transcription factors. Binding of cytokine to the receptor results in JAK-catalyzed phosphorylation of STAT proteins. Small-molecule inhibitors of JAK enzymes can block this signal.

An alternative strategy to intercepting the cytokine is to let it reach its destination but muffle its signal by inhibiting the protein kinase activity of the relevant member of the JAK family (Figure 17-1: note the label "JAK INHIBITOR DRUG"). A small number of compounds that accomplish this have reached the market, the prototype of which is Xeljanz® (generic name: tofacitinib.)[93] It inhibits the JAK family enzymes with relatively broad specificity and is approved to treat RA, the inflammatory skin condition psoriasis, and the bowel condition ulcerative colitis. Xeljanz® is often coadministered with a biologic, but achieved more than $1.7 billion of sales in 2018.

Xeljanz® was initially developed to control graft rejection following transplantation surgery. It has continued to be of interest to this area of medicine, but has not yet advanced to receiving FDA approval.[94] Like other immunosuppressive agents, its beneficial actions come with some risk of increasing the patient's susceptibility to infections. When a candidate drug has potential in several different areas of medicine – and this is increasingly common, as research uncovers the activity of targets that are close to central control mechanisms within cells – a pharmaceutical company must choose one area in which to pursue a first approval. It is logical to select the one in which commercial potential is greatest. Xeljanz® has now been joined in the market by a competitor, Olumiant® (baricitinib).

With rapid advances in deconvoluting the complex language of cytokine signaling, compounds that target specific JAK-family enzymes more specifically than tofacitinib are being developed by a number of companies for a variety of inflammatory conditions.

Finally, we will touch on a few of the other new classes of small-molecule drugs with the potential to command large markets that have come to market since 2000.

Kinase Inhibitors. As we have seen with the JAK family, protein kinases are officer-class enzymes that regulate the activities of other enzymes and proteins. They do this by transferring a phosphoryl group from ATP to their protein substrates, an action that is dynamically checked and balanced by protein phosphatases. Kinase action that has escaped from its proper control because of a mutation is often a driver of uncontrolled cell division in cancer. About 2% of human protein-coding genes encode protein kinases, of which there are about 560 in all.

Because all protein kinases use ATP as one of their substrates, it initially seemed that the development of specific inhibitors of particular kinases would be extremely challenging. Medicinal chemistry, greatly assisted by structural biology, has gloriously overcome this obstacle, and about 22 members of the kinase inhibitor class of drugs have received FDA approval (updated lists are readily available). Most of these are used to treat various types of cancer, but tofacitinib (discussed above) is used to control autoimmune responses and represents the first example of a kinase inhibitor moving beyond cancer therapy in human medicine.

Interestingly, a veterinary agent called Apoquel® (oclacitinib) is used in dogs to suppress the itching caused by allergic dermatitis. This compound works by inhibiting JAK1 and JAK3.

SGLT2 Inhibitors. The sodium/glucose cotransporter 2 (SGLT2) of the early proximal tubule is the major gateway for renal glucose reabsorption. In type 2 diabetes, when the body has an excess of circulating glucose, disposing of some of it through the urine becomes advantageous, and SGLT2 inhibitors permit this to happen. Four have reached the US market at the time of writing: Farxiga® (dapagliflozin: AstraZeneca), Jardiance® (empagliflozin: Boehringer Ingelheim), Invokana® (canagliflozin: Janssen Pharmaceuticals, a unit of Johnson & Johnson) and Steglatro® (ertugliflozin: Merck, codeveloped with Pfizer).

The flozins, or gliflozins, initially looked like a humdrum addition to the formulary of medications available to help manage type 2 diabetes. They have turned out to deliver unexpectedly strong benefits, including reductions in cardiovascular events.[95]

Hepatitis C Drugs. Unqualified use of the word "cure" is unusual in discussions of drug discovery, but it applies to antivirals developed to treat hepatitis C. Mavyret®, a combination therapy containing two agents that act at distinct viral targets, is marketed by AbbVie (North Chicago, IL) and is taken for eight weeks. Following its approval by FDA in 2017, favorable pricing relative to previously dominant medicines from Gilead Sciences (Foster City, CA) made Mavyret® the fastest-growing medicine in the sector.

The high cost of these agents has been controversial. Once again, the argument is made that an expensive drug taken for a limited time and likely to save the patient from even more costly complications represents a net savings to the health care system.

From a researcher's point of view, though, the fact that a chronic, and ultimately life-threatening disease can actually be cured is a confidence booster.

The End of the World is Nigh (Maybe)

18

In a world constantly disrupted by new technology, even the most successful company's dominance can melt away in a few years. Leaders of established biopharma companies know this, and are seldom either short-sighted or resistant to change. To the contrary, their consultant-driven compulsion to be early adopters of new ideas can amount to risky behavior that sometimes impresses Wall Street analysts more than experienced scientists.

Biopharma flourished from 1990-2010 by marketing blockbuster drugs to very large patient populations at moderate cost per patient. Statins to lower "bad" cholesterol, stomach-sparing cyclooxygenase-2 inhibitors for arthritic pain, and proton pump inhibitors for excess stomach acidity were all multibillion-dollar categories of drug in that era. Antihypertensives working by several different mechanisms were also big sellers, but the process of determining which type of agent worked best for any patient was one of trial and error, an inefficient process when a busy doctor might see the patient only at long intervals. With all of these widely prescribed agents, medicine was given on the basis of statistical evidence covering large populations rather than individual patients. Its benefit to any single patient had to be evaluated after the fact.

Blockbuster drugs – the ones that sell a billion dollars' worth or more each year – are highly acceptable when the market is enormous and the cost per patient can be kept small to encourage wide and continued use. The statins (Chapter 15) represent this model well, with some estimates placing the patient population near 20 million. Many such products introduced in the 1990's

Chapter 18: The End of the World is Nigh (Maybe)

have been losing patent protection in the decade since 2010, creating massive business pressure on biopharma companies. Today, still effective but providing minimal revenue to the companies that invented them, these medicines serve patients as low-cost generics. Rueful drug hunters describe that generation of medicines as "low-hanging fruit" that was *relatively* easy to ripen and pluck. Now the low branches are bare, and targets that offer simple routes to drug discovery are as plentiful as pineapples in a penguin colony.

One answer, of course, is to extend the ladder of science to reach higher branches. As we noted in Chapter 17, crystallographic structural analysis was recently made applicable to G protein-coupled receptors (GPCRs), a class of proteins that includes the targets of something like 40% of all known drugs. These receptors exhibit fiendishly subtle interactions with their ligands – agonists, antagonists, partial agonists, and so forth – and the capacity to visualize these interactions is a huge advance toward making them tractable.

A second profound rationalization has been acceptance that screening compound libraries against purified targets cannot supply enough success to feed the industry's need for new progress. The response is new interest in screening against intact biological systems (also discussed in Chapter 17). New technologies, especially those based on fusions of intrinsically fluorescent proteins with cellular components, provide the sophisticated readouts required.

Even the industry's most stalwart establishments pulsate with willingness to change, react and adapt to new challenges. Blissful ignorance is not the problem, nor is it one of lazy attitudes. The problem is that discovering new medicines is hard and getting harder.

On top of the target supply problem comes the fragmenting impact of new technology. Personalized medicine is on the way, but is arriving the way the Hudson River arrives in New York City – continuously, and for a very long time. As cancer is a set of widespread and life-threatening diseases of genetic malfunction, it is not surprising that the DNA era has delivered the most progress in oncology. For over a decade, for example, analyses of gene expression signatures in breast tumors have allowed patients to be placed in subgroups for which particular treatments give the best clinical outcomes. The information being assessed in this process continues to be refined, and gene-targeted or whole-genome sequencing is increasingly part of it. It is a landmark, for example, that the Center for Medicare and Medicaid Services announced on March 16, 2018, that it will cover a particular test that surveys 324 tumorigenesis-relevant genes and two genomic signatures by next generation sequencing for cancer patients that it insures.

There is also the case of the drug crizotinib, which inhibits an altered form of the enzyme anaplastic lymphoma kinase (ALK) that is created by a chromosomal rearrangement and drives the disease in 4% of cases of non-small cell lung carcinoma. Genetic testing of tumor biopsies allows the drug to be given only to patients likely to benefit from it, which makes sense medically and from the perspective of insurers. But there is a correlate: all cost recovery and profit now has to come from treating the selected patient group, which can be tiny. Instead of the millions of people who take statins, the number of new cases of ALK-positive non-small cell lung carcinoma in the U.S. per year is around 10,000.

In the limit of this trend, the condition of every patient is recognized as unique, and the physician's long-standing ambition to "treat the patient, not the disease" will, in a sense, be realized.[96] We are nearly at this point with some forms of cancer, but medical science aims to make it the norm for all pathologies. Clearly, this progress toward individualized diagnosis and treatment disrupts the paradigm based on huge patient pools under which biopharma

Chapter 18: The End of the World is Nigh (Maybe)

thrived just a short time ago. As patient pools are divided into smaller and smaller subgroups, the opportunities to recommend one drug for a vast population tend to be reduced.

The current remedy is to charge very high prices for drugs that treat small patient groups, especially those with rare and life-compromising genetic diseases. Fairly or not, exasperation with the industry is growing as society engages the problem of meeting its medical needs at a cost that it thinks it can afford. Particularly in countries that consider health care an entitlement that comes with citizenship, the argument that expensive drug treatment is still cheaper than more labor-intensive interventions is not calming the critics. Something will have to give.

Prevention, you cry? Yes, I agree, if people would abandon bad habits that increase their risks of disease, the world would be a healthier place. Bans on public smoking have surely been a good thing, and perhaps the move in some jurisdictions toward taxing foods with a high sugar content will also reduce the numbers of people suffering from metabolic diseases such as diabetes and obesity. We must hope so. But these excellent measures will not find a job for our hopeful Young Scientist.

Biopharma is doing everything it can to preserve its way of doing business while adapting to new challenges. A 2018 analysis of portfolios and pipelines predicts that prescription drug sales worldwide will grow more than 6% in the 2018-2024 period and that most major companies will achieve moderate growth through that period.[97] But there are those who argue that the entire business model of medicines in a bottle is outdated thinking, soon to be swept away by a new enlightenment based on phenomenally powerful and patient-specific technologies of early diagnosis and, indeed, pre-diagnostic predictive medicine.

The thinking Young Scientist may even be working hands-on with the new technologies that will launch the revolution that some expect. The question is, though, where is future science going to take biopharma? How does the Young Scientist select the right wave, or tiger, to ride?

I referred more than once to the reasonable but unfulfilled expectation that the Human Genome Project would uncover a vast number of potential drug targets that would create a bonanza for the industry. Its current equivalent is that genome wide association studies (GWAS) will uncover unsuspected connections between particular genes and certain disease states, so that persons carrying the disease-linked variant of a gene might be candidates for treatment with a suitable drug. The emerging problem is that many linkages can be found, but they are mostly weak. Also, and very importantly, widespread conditions that are huge factors in public health – hypertension, cardiovascular disease, and the like – tend to be governed by weak contributions from a large number of genes.

Strong linkages of disease to specific genetic variants certainly can be found. Persons who are homozygous for the ApoE4 variant of the gene encoding apolipoprotein E live with about 16 times higher risk of developing Alzheimer's disease than persons who are homozygous for the ApoE2 form. The two forms differ at two positions in the 299-residue mature protein. If we could find numerous "smoking guns" of the same kind, they would undoubtedly stimulate activity toward drug discovery. However, the current view of the value of GWAS seems to be of the "more work is needed" kind, and it may be that rather high-level conclusions are the main product for the foreseeable future. For example, it is reported in 2018 that asthma and allergic diseases have deep connections at the genetic level,[98] but this kind of information seems unlikely to provide much fuel to drug discovery other than, possibly, stimulating cross-testing of medicines between these two areas.

My personal view of GWAS likens it to astronomy, which

Chapter 18: The End of the World is Nigh (Maybe)

massively changed humanity's understanding of the universe over the past century. It was only in the 1920's that distant, misty objects in the night sky came to be recognized as galaxies beyond our own Milky Way. Today we are told that there are at least one hundred billion such galaxies. Other recent progress has proved as fact the inevitable existence of planetary systems surrounding stars other than our Sun. Given the profound complexity and diversity of our genetic make-up, it may take decades for advanced genomic analysis to pay off in comparable terms. The hoped-for difference, of course, is that medical knowledge will measurably improve the human condition in the most tangible way, while knowledge of the majesty of the universe will remain a source of intellectual satisfaction.

Another paradigm-busting development is therapies based on cells rather than medicines. The CAR-T approach to treating both blood cancers and solid tumors,[99] for example, has already recorded enough success to generate a full-scale scramble to respond to the opportunity. In this method, a cancer patient's T cells are removed and engineered to express a receptor that recognizes a tumor-specific antigen. They are then reinfused into the patient with the mission of attacking and disposing of the tumor cells. While it is costly, this approach clearly has a significant part to play in the future of oncology.

Gene therapy has knocked on medicine's door for quite some time, but the advent of CRISPR/Cas9 gene-editing methods and their variants shows every sign of being transformative. It remains to be shown that the method will be safe enough for widespread medical use, but there is hope that this is the rescue ship so urgently needed by patients affected by genetic diseases. Young Scientists with the proper training can join this revolution.

CRISPR/Cas9 may be the new superstar, but longer-running efforts toward successful gene therapy also have the capacity to break through. For both hemophilia A and hemophilia B, each of which is caused by lack of a specific clotting factor, gene therapies based on adenoviral vectors have advanced to Phase 3 clinical trials. Patients receiving the treatments were largely able to discontinue injections of recombinant clotting factors. Note the two-edged nature of advances like this. They will have winners and losers. Patients, hopefully, and the inventors and suppliers of the new therapies will gain, but makers of the displaced therapies will lose market share.

In any gold rush, we are often told, the makers of shovels and wheelbarrows get rich while most of the gold miners limp away with broken dreams. The world of biopharma may be a little like this, in that some of the hottest companies supply information and enabling technology to others rather than engaging directly in discovery. Do not overlook the opportunities for bright Young Scientists in this space.

Here are some examples, chosen to some extent for diversity, of smaller companies attempting to bring truly pioneering science to the commercial sphere. Rather than suggesting that you seek out a job with one of them, I want to highlight the excitement to be found in smaller, adventurous companies that you may find it hard to match in a larger biopharma. Which arena suits you best is up to you.

We are interested in the employment of scientists in the United States, so this list is confined to American-based companies.

Grail (Menlo Park, CA). This company's mission is to detect cancer much earlier than has been typically possible by detecting and sequencing circulating tumor DNA. The challenge is to recognize faint tumor-derived signals against a high background.

Aileron Therapeutics (Cambridge, MA). Aileron attempts to bridge the gap between small-molecule therapeutics and large molecules by investigating the capabilities of conformationally-stabilized "stapled" peptides to enter cells and disrupt protein-protein interactions.

Wave Life Sciences (Cambridge, MA). Proprietary chemistry is applied to making improved "stereopure" nucleic acid-based therapeutics with potential to benefit patients affected by rare diseases. Wave Life Sciences and Aileron are both based on technologies originated by Professor Gregory Verdine of Harvard University.

Denali Therapeutics (South San Francisco, CA) Seeking new therapeutics for neurodegenerative diseases.

Moderna Therapeutics (Cambridge, MA). Uses modified messenger RNA molecules (note "rna" in its name) as drugs that induce the production of encoded proteins with a view to changing cell behavior.

Arvinas (New Haven, CT). One of several companies exploring the potential of "protein degraders," two-headed molecules that bring a targeted protein molecule into contact with the cell's protein degradation machinery.[100] The goal is to induce the destruction of a protein responsible for driving a disease. Among several other small companies operating in this space are **Kymera Therapeutics** (Cambridge, MA) and **C4 Therapeutics** (Watertown, MA), while there has also been significant interest inside major biopharma companies.

Abide Therapeutics (San Diego, CA). Targets members of the serine hydrolase family of enzymes using the tool of activity-based protein profiling. Like many companies mentioned in this list, Abide has an exceptionally distinguished Scientific Advisory Board and is an offshoot from innovative academic work.

Relay Therapeutics (Cambridge, MA). Strong interest has emerged in finding drugs that bind to their protein targets allosterically, i.e. at sites remote from the active site. This mechanism may provide a way to alter the activity of previously difficult targets. Relay Therapeutics studies protein dynamics (movement) using high-end computing power to open new routes to drug discovery.

BlackThorn Therapeutics (San Francisco, CA). Self-described as a clinical-stage biopharma endeavoring to refine physiological correlates of complex neurobehavioral diseases to optimize patient selection for trials of targeted candidate agents.

Aravive (Houston, TX). Developing a range of soluble forms of the AXL receptor to intercept its ligand Gas6 as a means to suppress metastasis in breast cancer and ovarian cancer. High natural affinity (K_d near 30 pM) between the ligand and receptor has driven Aravive to develop soluble receptors with *subpicomolar* affinity for Gas6.

Seattle Genetics (Bothell, WA). A leading center of expertise in antibody-drug conjugates, therapeutic molecules composed of a potent cytotoxin covalently linked to a monoclonal antibody with high affinity for a tumor-specific antigen. Antibody-mediated delivery of the toxic payload to the tumor is intended to cause the tumor to be eliminated. **ImmunoGen** (Waltham, MA) is another longtime participant in the same field.

Exicure (Skokie, IL). This company explores the therapeutic potential of "spherical nucleic acids" (SNA^{TM}), oligonucleotide-coated nanoparticles that undergo receptor-mediated uptake into cells of many different types. Selection of the oligonucleotide sequence allows different effects to be evoked in the target cells, including immunomodulatory actions.

Caribou Biosciences (Berkeley, CA), **CRISPR Therapeutics** (Cambridge, MA), **Editas Medicine** (Cambridge, MA), **Intellia Therapeutics** (Cambridge, MA), **Arbor Biotechnologies**

(Cambridge, MA) and **Beam Therapeutics** (Cambridge, MA) are all companies founded in the past few years to pursue the immense promise of CRISPR/Cas9-based methods of gene editing. The legal battle over intellectual property rights to the technology is expected to be long, and could have dramatically positive or negative consequences for some of these companies.

Mammoth Biosciences (San Francisco, CA) is another CRISPR/Cas9 technology-based company, but specializes in diagnostics. The sequence-finding and -cutting capacity of the method is equipped with a fluorescence-based readout to provide rapid detection of specific DNA sequences suspected to be present.

eFFECTOR Therapeutics (San Diego, CA). eFFECTOR is self-described as a clinical-stage company working to develop the potential of selective translation regulators as anticancer agents. These compounds are intended to block the production of particular proteins by interfering selectively with translation of the mRNA that encodes them.

Forge Therapeutics (San Diego, CA). Takes a new approach to the classical problem of inhibiting metalloenzymes with a view to developing novel antibiotics.

19 If It Feels Good, Do It!

The song title that gives this closing chapter its title aligns poorly with the dedication and self-discipline that the scientific life requires and rewards.[101] Even so, it sums up the final effort of will that some of us may still need to make to escape the old idea that working in industry is somehow a less elevated calling than staying the course in "pure" research.

For one thing, the notion of purity in academic research is a fiction. Career pressures, the personal flaws (including your own) of everyone involved, growing economic constraints and the determination of many scientists to compete ruthlessly for the rewards on offer make the academic scene challenging for the idealistic researcher who prizes truth (as science finds it) over personal advancement. I am not damning the world of basic research or calling it a swamp that needs to be drained, but it is no longer – if it ever was – a place for the modest, the faint-hearted or the timid.

Consider the governing forces that rule the world of industrial research and, specifically, that of biopharma. Working in industry, you must aim to make medicines so safe and effective that you would, on medical advice, administer them to your own children. The patient, in any event, is always somebody's child, parent, sister or brother – a precious part of the human family. My experience has been that the huge majority of people working in the industry keep that reality constantly in mind, and make it the guiding principle of their work.

Additional fences surround industrial work to keep it within the bounds of correct science. The costs of drug development are

Chapter 19: If It Feels Good, Do It!

enormous, and they grow as a project moves from its preclinical beginnings into its later stages. If shaky data are accepted as a basis for progressing a program beyond its early phase, the inevitable result is that the project will crash and burn at a later stage when far more resource has been consumed than if it had been terminated early on. Moreover, another program with a sounder basis may have been deemphasized when the faulty one received support. All of this adds up to a need for absolute rigor to be enforced in company labs. Slackness will be a costly luxury.

I can say with full sincerity that I was never, in 32 years of hands-on research in biopharma, asked in even the slightest way to provide anything other than completely sound and objective scientific data. Nor did I see any colleague asked to lower their standards on any occasion.

The main reason for this was the integrity of all the individuals involved and the company's insistence on a culture that upheld it. Beyond this was the cold economics of the process. A successful drug brings major economic rewards, and it is far more logical for companies to pursue this legitimate prize than to engage in conspiracy or deception with a view to fooling physicians and regulators. Bad behavior, quite clearly, can occur in biopharma, but it has more chance of achieving some ill-gotten gains in the realms of marketing and promotion. The nature of biopharma's internal scientific process is such that the highest standards are also the best business practice.

This is why I have reached for this chapter's title. For the researcher who dearly wishes to perform excellent research that falls or stands on its merits rather than on his or her capacity to self-promote, biopharma can most definitely provide a wonderful environment for self-actualization. This book has been about encountering the industry with open eyes, but it is not intended to talk you out of working in it. Quite the reverse.

Whichever company you work for, there are some general rules of how to survive and prosper in an ecosystem in which there is constant oversight and evaluation of every individual's professional value.

Be yourself. The attribution is shaky, but the Irish playwright and wit Oscar Wilde supposedly said "Be yourself; everyone else is already taken." True enough, whoever said it. Successful company workers express their individualism while staying within the company's accepted range of styles and behaviors. Just remember that your greatest strength will often be your greatest weakness. For example, if you are scientifically cautious and careful (I plead guilty), you will sometimes deliver conclusions more slowly than your customers would prefer. Also be aware that it's the weakness that you are not aware of that will hurt you most: you can do something about the ones that you recognize.

Do what you do well. Talents differ, and different people operate best in different roles. Supervising other people is a serious responsibility and *should* be stressful to some extent, but if you find it impossible (even after training) to deliver occasional difficult messages, then you should probably not chase a position on your company's management ladder. Recognize too that an inability to deliver positive messages is even more of a handicap.

Nearly everybody faces a challenge when handling authority. Some fail to understand the distinction between authority and power, and become authoritarian. Others shrink from the harder duties that organizational leadership entails. Finding the proper balance is a challenge. When your manager has achieved a perfect balance of the carrot and the stick, you are being managed well.

Always do research the right way. Company research is usually applied research with a project goal behind it, and understood to be different in kind from basic or pure research. And, of course, real differences exist. The wise academic researcher selects an

Chapter 19: If It Feels Good, Do It!

important problem that seems likely to crack after appropriate effort. A classic example is that of Christian Anfinsen, the pioneer of protein folding studies. He did not select ribonuclease as his protein of interest and then decide to see whether it could refold from a denatured state. He found a protein – ribonuclease – that was able to refold, then set about discovering how it did so.[102] A drug discovery team does not have this luxury. If a certain protein is a good drug target, then that this is the protein that must be studied, however gnarly its properties.

Despite this variation, research is always research and can only be tackled properly in one way. Every experiment must begin with what you know, and take one step into what you don't know. Preferably, the experiment should have two possible outcomes, and you will know in advance what each of them will mean.

Once in a while, the result will leave you shocked – *how could that possibly happen?* Sometimes, of course, there has been a mistake or miscalculation, and order is restored when you retrace your steps. But when that's not the case, and you genuinely are boggled by what you found, then it's time to be happy. Pretty soon, you will have learned something.

It is widely understood that solving a trivial problem in science can take just as much effort as solving an important one. This is why basic researchers should strive to get involved with intensely significant questions. The applied researcher may have to apply just as much intellectual force to his or her supposedly trivial problem, but the work is validated by its overarching purpose of seeking a new medicine and by the way in which it fits into a larger effort.

Do what you do better than anybody else in the company. Perhaps surprisingly, the challenge in a science-based company has more to do with bridging the gap between different people's expertise than with conflict between people whose skills overlap. Inevitably though, and especially in big companies with multiple research centers, you will know and sometimes work with fellow employees who operate in the same general area as you do. You need to assess your own performance level and that of your team against the level of your colleagues. If you feel that they outshine you, whatever the reason, it's time to be concerned. On any occasion that leads to cutbacks, you and your unit will be at risk.

Always do the right thing. If that doesn't work, you are in the wrong company anyway. Integrity is the essence of the biopharma, or "ethical pharmaceutical" industry, and integrity starts with you. If you drop the tube containing a week's work (why do I keep mentioning this?), tell your boss directly and live with the consequence. If the results don't fit the pattern that was hoped for, again, tell the story straight and do not even think of "fixing" the data. If you see a violation of company standards elsewhere, go to the proper person and rely on the company's values to see that things are corrected. If doing any of the above leads to unreasonable negativity for you (and it is best not to drop the tube every week,) then you are in an organization with poor leadership and standards, and should move on when you can. Your purpose in being at a company is to have a satisfying career *while helping to bring great new medicines to patients.* Bad science and bad behavior make both impossible.

Work safely. A small number of tragic accidents in academic laboratories have led to a new emphasis on proper standards.[103] Industrial labs, in contrast, have had the benefit of relatively strict oversight by federal and state agencies, and are widely regarded as doing better. Despite this, safe work is like safe driving, a matter of everyday caution and good habits. If you see somebody letting standards slip by failing to wear eye protection or handling a chemical incorrectly, you need to find a way to correct the situation without causing undue friction. Saying that "it's not my responsibility" is not good enough.

Stay in shape. Maintaining personal physical fitness is one of the biggest favors that you can do for yourself and your career. Different bodies require different regimes to stay in tune, but the hard work required of a scientist will be easier to sustain if you protect your energy level and capacity to get through long days by maintaining fitness.

Constantly check that your goals and the company's goals are aligned. Your links to the company need to work for both parties. Working for a company means exactly what it says – in transactional terms, you really do work *for* the company and not for yourself. The key to satisfaction is that the work also satisfies your own need to be challenged, to demonstrate your talent, to explore the amazing vistas that are constantly revealed at the frontiers of science, and to make you feel that you are living the life of a scientist in a way that adds value to science as a whole.

Acknowledgements

This book began as a two-day course presented to students in the BIOTrain program at Montgomery College in Germantown, Maryland, in December 2017 and January 2018. I am grateful to Dr. Mike Smith, coordinator of BIOTrain at that time, for inviting me to present the course, and admired the commitment and enthusiasm of the students and professionals who attended two long Saturday sessions.

Jim Boyd, Bill Curatolo, Alison Hackett, John LaMattina, Joanne Murray, Bess Sorensen, Chris Southgate and Ian Williams kindly read drafts and made constructive suggestions, many of which I adopted. The views expressed remain my own, as do any errors that the reader may detect.

Kieran Geoghegan

2019

ENDNOTES

[1] P. B. Medawar, *Advice to a Young Scientist*, Basic Books, 1981 (revised ed.).

[2] E. O. Wilson, *Letters to a Young Scientist*, Liveright, 2013.

[3] Wilson was not demeaning heavily populated fields. He meant to remind the reader that the details of seemingly mundane work can harbor the chance of an unexpected but significant discovery. In protein chemistry, the area in which I worked, interesting anomalies cropped up often and looking into them was rewarding and enjoyable.

[4] After ingesting a culture of *Helicobacter pylori* in 1984, Marshall developed a form of gastritis that can progress to stomach ulcers. Accounts of this self-experiment are readily found online.

[5] B. Marshall, L. Hendry and B. Caleo, *How to Win a Nobel Prize*, Piccolo Nero, 2018.

[6] J. M. Bishop, *How to Win the Nobel Prize: An Unexpected Life in Science*, Harvard University Press, 2004.

[7] P. Doherty, *The Beginner's Guide to Winning the Nobel Prize: Advice for Young Scientists*, Columbia University Press, 2008.

[8] P. J. Feibelman, *A PhD is Not Enough!*, Basic Books, 2011.

[9] J. D. Watson, *The Double Helix*, 1968. Seek out the Norton Critical Edition (ed. G. S. Stent, 1980), in which the text by Watson is bundled with the original scientific publications and fascinating book reviews by prominent figures.

[10] This put-down – "A modest man, who has much to be modest about" – was most famously applied by Winston Churchill to Clement Atlee, who succeeded him in 1945 as British Prime Minister.

[11] For example, see K. Powell (2018) Industrial Strength *Nature* 555, 549-551 (doi: 10.1038/d41586-018-03306-1). Available online.

[12] For a brief discussion, see the summary of a Consensus Study Report of the National Academies of Sciences, Engineering and Medicine (2018) entitled "The Next Generation of Biomedical and Behavioral Sciences Researchers" at https://www.nap.edu/resource/25008/NextGenerationReportHighlights.pdf

[13] For information about recent annual budgets of the National Institutes of Health, browse to www.hhs.gov, select the A-Z index, locate the Budget section, and select the subsection for NIH.

[14] Browse to https://www.nsf.gov/ and refer to section on Funding.

[15] Source: Centers for Medicare and Medicaid Services.

[16] J. D. Watson, A. Berry and K. Davies, *DNA*, A. A. Knopf, 2017, page 169.

[17] Source: Data table for Figure 17 of *Health, United States, 2017* (web-accessible at https://www.cdc.gov/nchs/data/hus/hus17.pdf from the Centers for Disease Control and Prevention).

[18] The industry's collective voice is the Pharmaceutical Research and Manufacturers of America (PhRMA) (https://www.phrma.org/). It articulates the consensus of research-based companies on matters such as pricing, patent law, drug imports and Medicare policy. Opposing opinions can often be found in the media, or in the comments of advocacy groups such as Public Citizen (https://www.citizen.org/).

[19] A debate continues between cardiologists over the use of stents to treat patients with stable ischemic heart disease (serious blood vessel blockage without an acute heart attack). See S. P. Sedlis et al. (2015) Effect of PCI on Long-Term Survival in Patients with Stable Ischemic Heart Disease *New Eng. J. Med.* 373:1937-1946 (doi: 10.1056/NEJMoa1505532). A clinical trial comparing outcomes between patients receiving stents and patients treated with medication and lifestyle change showed no advantage for the expensive surgical approach. Some commentators, easily found online, suggest that the financial gain to physicians associated with the use of stents has created reluctance to act upon this result.

[20] The evolution of research strategies in biopharma companies is the subject of much commentary. Examples available online include (i) A. Guatam and X. Pan (2016) The changing model of big pharma: impact of key trends, *Drug. Discov. Today* 21, 379-384 (doi: 10.1016/j.drudis.2015.10.002); (ii) A. Schuhmacher, O. Gassmann and M. Hinder (2016) Changing R&D models in research-based pharmaceutical companies *J. Transl. Med.* 14, 105 (doi: 10.1186/s12967-016-0838-4); (iii) H. Naci, A. Carter, and E. Mossialos (2015) Why the drug development pipeline is not delivering better medicines *BMJ* 2015;351:h5542 (accepted version available at http://eprints.lse.ac.uk/64220/).

[21] P. Cullen, Cystic fibrosis drugs to cost €650m over 10 years, *The Irish Times*, May 5, 2017. Available online.

[22] NICEimpact cancer. Published January 2018 by the National Institute for Health and Care Excellence of the United Kingdom. Available online at https://www.nice.org.uk/media/default/about/what-we-do/into-practice/measuring-uptake/nice-impact-cancer.pdf.

[23] Source: company Annual Reports.

[24] For a nice one-page summary of the FDA's Drug Review Process, browse to www.fda.gov and enter "ensuring drugs are safe and effective" in the Search box. Select the top hit from the search, headed The FDA s Drug Review Process: Ensuring Drugs Are Safe and Effective.

[25] Search online for *Statins: A Success Story Involving FDA, Academia and Industry* by S. W. Junod, originally published in the March/April 2007 issue of *Update*, a publication of the Food and Drug Law Institute, Washington, DC.

[26] "Statement from FDA Commissioner Scott Gottlieb, M.D. on agency's efforts to advance development of gene therapies," July 11, 2018. This is online at https://www.fda.gov/NewsEvents/Newsroom/PressAnnouncements/ucm613026.htm, and contains links to six documents presenting draft guidances and guidelines related to specific areas, including hemophilia and retinal diseases.

[27] To learn about medical writing, visit the web site of the American Medical Writers Association (AMWA) at https://www.amwa.org/.

[28] WIPO has an excellent FAQ page on patents worldwide at http://www.wipo.int/patents/en/faq_patents.html.

[29] Quotation from Annie Dillard, *The Writing Life*, HarperCollins, 1989.

[30] A. H. Maslow, *A Theory of Human Motivation*, 1943: the central idea is widely discussed, e.g. at en.wikipedia.org.

[31] For an illustrated introduction, see the website of ThermoFisher Scientific Electron Microscopy Solutions at https://www.fei.com/life-sciences/history-of-cryo-em/.

[32] R. Aebersold and M. Mann (2016) Mass-spectrometric exploration of proteome structure and function, *Nature* 537, 347–355 (doi: 10.1038/nature19949). Paywall-protected.

[33] For engaging wisdom on excellence in medicinal chemical design and synthesis, see M. A. Murcko (2018) What Makes a Great Medicinal Chemist? A Personal Perspective, *J. Med. Chem.* J. Med. Chem. 61, 7419-7424 (doi: 10.1021/acs.jmedchem.7b01445).

[34] For information about Good Manufacturing Practice, visit the web site of the International Society for Pharmaceutical Engineering (www.ispe.org), which has a section titled GMP Resources.

[35] *Nature* 506, 131–132 (2014) (doi:10.1038/506131b)

[36] P. J. Feibelman, ibid. pp. 101-105.

[37] G. T. Doran (1981) There's a S.M.A.R.T. Way to Write Management's Goals and Objectives, *Management Review* 70, 35-36.

[38] For a crisp defense of this system, see a blog by the executive most identified with it: "Jack Welch: 'Rank-and-Yank'? That's Not How It's Done," website of *The Wall Street Journal*, Nov. 14, 2013. Rejecting its blunter aspects does not equate to rejecting all performance reviews. For the counter-argument to Welch, see a blog by Stephen R. Satterwhite: "Here's To The Death Of Microsoft's Rank-And-Yank," website of *Forbes*, Nov. 13, 2013.

[39] For an account of the origins of this song, see http://buffysainte-marie.com/?p=809

[40] X. Hua, N. Carvalho, M. Tew, E. S. Huang, W. H. Herman and P. Clarke (2016) JAMA. 315, 1400-1402 (doi:10.1001/jama.2016.0126).

[41] J. LaMattina, *Devalued and Distrusted: Can the Pharmaceutical Industry Restore its Broken Image?*, Wiley, 2012, pp. 105-106.

[42] L. M. Jarvis (2018) Civica Rx aims to shake up the generics market. C&E News 96, issue 48, Dec. 3, 2018.

[43] Browse to http://www.icmje.org/ and select Recommendations.

[44] S. S. Hughes, *Genentech: The Beginnings of Biotech*, University of Chicago Press, 2011.

[45] Text taken from the Laboratory's web site. Information about it and others referred to in the chapter can be found by routine internet searching.

[46] J. LaMattina (2012) The NIH Is Going to Discover Drugs... Really?, *Forbes*, available online.

[47] The web site of the Academic Drug Discovery Consortium (ADDC) is at http://addconsortium.org/

[48] E. Martens and A. L. Demain (2017) The antibiotic resistance crisis, with a focus on the United States, *J. Antibiot.* 70, 520–526 (doi: 10.1038/ja.2017.30). Available online.

[49] L. L. Ling et al. (2015) A new antibiotic kills pathogens without detectable resistance, *Nature* 517, 455–459 (doi:10.1038/nature14098). Paywall-protected.

[50] A. Parmar et al. (2018) Design and Syntheses of Highly Potent Teixobactin Analogues against *Staphylococcus aureus*, Methicillin-Resistant *Staphylococcus aureus* (MRSA), and Vancomycin-Resistant Enterococci (VRE) *in Vitro* and *in Vivo*, *J. Med. Chem.* 61, 2009–2017 (doi: 10.1021/acs.jmedchem.7b01634). Paywall-protected.

[51] J. Park et al. (2017) Plasticity, dynamics, and inhibition of emerging tetracycline resistance enzymes, *Nat. Chem. Biol.* 13, 730–736 (doi: 10.1038/nchembio.2376). Paywall-protected.

[52] S. M. Drawz and R. A. Bonomo (2010) Three Decades of β-Lactamase Inhibitors, *Clin. Microbiol. Rev.* 23, 160–201 (doi: 10.1128/CMR.00037-09). Available online.

[53] Access to Medicine Foundation (2019), "Are pharmaceutical companies making progress when it comes to global health?" Available online at the Foundation's website.

[54] A. D. Hopkins (2005) Ivermectin and onchocerciasis: is it all solved? *Eye* 19, 1057–1066 (doi:10.1038/sj.eye.6701962). Available online.

[55] A. W. Solomon et al. (2004) Mass Treatment with Single-Dose Azithromycin for Trachoma, *New Engl. J. Med.* 351, 1962-1971 (doi: 10.1056/NEJMoa040979). For a news report of remarkable progress, see D. G. McNeil, Jr., Now in Sight: Success Against an Infection That Blinds, *New York Times*, July 16, 2018. Both articles are available online.

[56] J. LaMattina (2015) The Role Of Big Pharma In Neglected Diseases, *Forbes*, available online.

[57] More information is available from the Centers for Disease Control and Prevention at https://www.cdc.gov/parasites/chagas/biology.html

[58] For a brilliant account of the items related to Groves and Oppenheimer as well as the entire Manhattan Project and its scientific antecedents, read Richard Rhodes, *The Making of the Atomic Bomb*, Simon and Schuster, 1986.

[59] J. Kunetka, *The General and the Genius*, Regnery History, 2015, p. 164.

[60] S. A. Blumberg and G. Owens, *Energy and Conflict: The Life and Times of Edward Teller*, G. P. Putnam's Sons, First Edition (1976). On page 128, the authors write: "The evidence strongly suggests that Teller was somewhat of a malcontent the very day he started working at Los Alamos."

[61] P. Morgan et al. (2012) Can the flow of medicines be improved? Fundamental pharmacokinetic and pharmacological principles toward improving Phase II survival, *Drug Discov. Today* 17, 419-424 (doi: 10.1016/j.drudis.2011.12.020). Paywall-protected.

[62] For a neat statement of this argument, see M. Rosenblatt (2014) The Real Cost of "High-Priced" Drugs, Harvard Business Review, online at https://hbr.org/2014/11/the-real-cost-of-high-priced-drugs. Dr. Rosenblatt was formerly the Chief Medical Officer of Merck and Co.

[63] See INN Working Document 09.251 (2009) *General policies for monoclonal antibodies*. Available online.

[64] The International Trade Administration report containing this quotation is available online: search for "2016 Top Markets Report Pharmaceuticals."

[65] S. S. Hughes, *ibid*.

[66] L. Marks (2012) The birth pangs of monoclonal antibody therapeutics, *mAbs* 4, 403-412, (doi: 10.4161/mabs.19909).

[67] For a comprehensive and easy-to-read history of molecular biology and its many applications to biomedical advances, see J. D. Watson, A. Berry and K. Davies, *DNA*, A. A. Knopf, 2017.

[68] Source: QuintilesIMS report *Disruption and maturity: The next phase of biologics*. Authors: D. Kent, S. Rickwood and S. Di Biase. Available online. Describes market shares up to 2016.

[69] For a gripping read on this subject, see T. Hamilton and D. Coyle (2012) *The Secret Race*, Bantam Books, 2012. In the first edition, beware of occasional errors in references to the science of EPO and its action; EPO is made in the kidneys and stimulates red blood cell production in the bone marrow, but the book gives incorrect information. Also, glycerol, not glycol, is used as the cryoprotectant when blood is frozen.

[70] A. Tojo and S. Kinugasa (2012) Mechanisms of glomerular albumin filtration and tubular reabsorption *Int. J. Nephrol.* 2012, article ID 481520 (doi:10.1155/2012/481520).

[71] L. A. Ridgley et al. (2018) What are the dominant cytokines in early rheumatoid arthritis? *Curr. Opin. Rheumatol.* 30, 207–214 (doi: 10.1097/BOR.0000000000000470). Available online.

[72] K. Peppel et al. (1991) A tumor necrosis factor (TNF) receptor-IgG heavy chain chimeric protein as a bivalent antagonist of TNF activity, *J. Exp. Med.* 174, 1483–1489 (doi:10.1084/jem.174.6.1483).

[73] P. M. Ridker et al. (2017) Antiinflammatory Therapy with Canakinumab for Atherosclerotic Disease, *New Eng. J. Med.* 377, 1119-1131. Freely available online (doi: 10.1056/NEJMoa1707914).

[74] For an excellent short history, see P. Nair (2013) Brown and Goldstein: The Cholesterol Chronicles, *Proc. Natl. Acad. Sci. USA* 110, 14829-14832. Freely available online (doi: 10.1073/pnas.1315180110).

[75] J. C. Cohen et al. (2006) Sequence Variations in PCSK9 Low LDL, and Protection against Coronary Heart Disease, *New Eng. J. Med.* 354, 1264-1272. Freely available online (doi: 10.1056/NEJMoa054013).

[76] For a popular account of Folkman's work, see R. Cooke, *Dr. Folkman's War*, Random House, 2001.

[77] For an amusingly written but wholly serious review of the mammalian immune system, see K. M. Yatim and F. G. Lakkis (2015) A Brief Journey through the Immune System, *Clin. J. Am. Soc. Nephrol.* 10, 1274–1281 (doi: 10.2215/CJN.10031014).

[78] For a dramatic and instructive account of the discovery of Keytruda®, see David Shaywitz's article "The Startling History Behind Merck's New Cancer Blockbuster" available online from *Forbes*, July 26, 2017.

[79] C. Robert et al. (2015) Pembrolizumab versus Ipilimumab in Advanced Melanoma, *New Eng. J. Med.* 372, 2521-2532. Freely available online (doi: 10.1056/NEJMoa1503093).

[80] S. S. Hall, *A Commotion in the Blood*, Henry Holt, 1997.

[81] For two examples of comments along these lines, see (i) Editorial (2010) *Nature* 464, 649-650, and (ii) "The great DNA letdown", *Fortune*, April 8, 2010. Both articles are available online.

[82] For news coverage of this ongoing issue, see C. Willyard (2018) Expanded human gene tally reignites debate, *Nature*, 558, 354-355 (doi: 10.1038/d41586-018-05462-w). Available online.

[83] A. J. Hopkins and C. R. Groom (2002) The druggable genome, *Nat. Rev. Drug Discov.* 1, 727-730. (doi: 10.1038/nrd892). Paywall-protected.

[84] For a commentary on this, see 2010 CMR International Pharmaceutical R&D Handbook, Thomson Reuters, executive summary available online at http://ips.clarivate.com/m/pdfs/CMR-Factbook_Exec_Sum.pdf.

[85] M. Gittelman (2016) The revolution re-visited: Clinical and genetics research paradigms and the productivity paradox in drug discovery *Res. Pol.* 45, 1570-1585. Freely available online (doi: 10.1016/j.respol.2016.01.007).

[86] D. C. Swinney and J. Anthony (2011) How were new medicines discovered? *Nat. Rev. Drug Discov.* 10, 507-519.(doi: 10.1038/nrd3480). Paywall-protected.

[87] J. Eder, R. Sedrani and C. Wiesmann (2014) The discovery of first-in-class drugs: origins and evolution. *Nat. Rev. Drug Discov.* 13, 577-587 (doi: 10.1038/nrd4336). Paywall-protected.

[88] R. I. Aminov (2010) A brief history of the antibiotic era: lessons learned and challenges for the future, *Front. Microbiol.* doi: 10.3389/fmicb.2010.00134. Available online only, and cited by digital object identifier (doi).

[89] C. Ferec and G. R. Cutting (2012) Assessing the Disease-Liability of Mutations in CFTR, *Cold Spring Harb. Perspect. Med.* 2012;2:a009480 (doi: 10.1101/cshperspect.a009480). Available online.

[90] See *The Washington Post*, July 2, 2015 for an article by Brady Dennis entitled "Are risks worth the rewards when nonprofits act like venture capitalists?"

[91] S. Jayaraman et al. (2000) Mechanism and cellular applications of a green fluorescent protein-based halide sensor, *J. Biol. Chem.* 275, 6047-6050. (doi: 10.1074/jbc.275.9.6047). Available online.

[92] Vertex Pharmaceuticals provides information about these medicines at www.kalydeco.com, www.orkambi.com, and www.symdeko.com.

[93] M. E. Flanagan et al. (2014) Case History: Xeljanz™ (Tofacitinib Citrate), a First-in-Class Janus Kinase Inhibitor for the Treatment of Rheumatoid Arthritis, *Annu. Rep. Med. Chem.* 49, 399-416 (doi: 10.1016/B978-0-12-800167-7.00025-0.) Paywall-protected.

[94] F. Vincenti et al. (2015) Evaluation of the Effect of Tofacitinib Exposure on Outcomes in Kidney Transplant Patients *Am. J. Transplant.* 15, 1644–1653 (doi: 10.1111/ajt.13181). Available online.

[95] B. Zinman wt al. (2015) Empagliflozin, Cardiovascular Outcomes, and Mortality in Type 2 Diabetes, *New Eng. J. Med.* 373, 2117-2128. Freely available online (doi: 10.1056/NEJMoa1504720).

[96] "The good physician treats the disease; the great physician treats the patient who has the disease." – William Osler, Canadian physician, 1849-1919.

[97] EvaluatePharma, World Preview 2018, Outlook to 2024. Available online at http://info.evaluategroup.com/rs/607-YGS-364/images/WP2018.pdf.

[98] Z. Zhu et al. (2018) A genome-wide cross-trait analysis from UK Biobank highlights the shared genetic architecture of asthma and allergic diseases, *Nat. Genet.* (doi:10.1038/s41588-018-0121-0). Paywall-protected.

[99] At the time of writing (2018), a nice presentation of the CAR-T approach appears at the web site of the National Cancer Institute at https://www.cancer.gov/about-cancer/treatment/research/car-t-cells.

[100] M. Scudellari (2019) The Protein Slayers, *Nature* 567, 298-300 (doi: 10.1038/d41586-019-00879-3).

[101] *If It Feels Good, Do It* is a song written by Michael Vale and recorded in 1972 by Della Reese on H&L Records. It should not be confused with the song *If It Feels Good Do It* by the Canadian rock band Sloan, recorded in 2001.

[102] U.S. National Library of Medicine. The Christian B. Anfinsen Papers. Protein Folding and the Thermodynamic Hypothesis, 1950-1962. Available online at https://profiles.nlm.nih.gov/ps/retrieve/Narrative/KK/p-nid/16

[103] B. L. Bendieff (2016) Warning: This Lab May Cause Injury or Death, available online at https://undark.org/article/lab-safety-universities-oversight/.

ABOUT THE AUTHOR

Kieran Geoghegan is originally from Dublin, Ireland, where he graduated in Biochemistry from University College, Dublin. He received his Ph.D. from Cambridge University, and was subsequently a post-doctoral fellow at the University of California, Davis and Harvard Medical School. He worked in drug discovery at a major pharmaceutical company in Connecticut from 1984 until his retirement in 2017, and has authored or coauthored more than 100 scientific publications. He lives in Mystic, Connecticut, and can be reached at kgeoghegan36@gmail.com.

www.ingramcontent.com/pod-product-compliance
Lightning Source LLC
Chambersburg PA
CBHW071540220526
45469CB00003B/860